Hothouse Earth Extinction

Gigaton Carbon Dioxide Removal Now

Also by Thomas F. Valone

Zero Point Energy: The Fuel of the Future

Nikola Tesla's Energy Unplugged

Harnessing the Wheelwork of Energy

Electrogravitics Vol 1 and 2

Homopolar Handbook

The Future of Energy: Challenges, Perspectives, Solutions

The Zinsser Effect: Electrogravity

Bioelectromagnetic Healing: A Rationale for Its Use

Modern Meditation Technique

Practical Conversion of Zero Point Energy

The Future of Energy: Challenges, Perspectives, and Solutions

HOTHOUSE EARTH EXTINCTION

Gigaton Carbon Dioxide Removal Now

Thomas F. Valone, PhD, PE

JT Pro Investments, LLC

Hothouse Earth Extinction

Gigaton Carbon Dioxide Removal Now

Thomas F. Valone, PhD, PE

1st Edition - March, 2025

2nd Edition – April, 2025

ISBN: 9798312494457

JT Pro Investments, LLC
5020 Sunnyside Avenue, Suite 216
Beltsville MD 20705

A portion of the proceeds is donated to
Integrity Research Institute
www.Integrity-Research.org
Request a free IRI Publication Catalog sent by mail

Excerpts from this book may be used without permission with a citation of its source

The sale of this book benefits Integrity Research Institute,
a nonprofit charitable 501(c)3 research organization.
Tax deductible donations are also accepted through Network for Good
service accessible from IRI homepage.
Subscribe to **Future Energy eNews**, a monthly free service through
Constant Contact, also linked to the IRI homepage.
You may also download a PDF copy of the IRI **Publication Catalog**

For James Hansen

The world's greatest climatologist

A **HOTHOUSE** IS A PLEASANT GREENHOUSE THAT HAS TRAPPED SOLAR RADIATION AND BECOME TOO HOT.

THINK **VENUS**, A PLANET THE SAME SIZE AS EARTH AND NOT REALLY MUCH CLOSER TO THE SUN. HOWEVER, VENUS HAS *ACCUMULATED TOO MUCH CO2*, WHICH TRAPPED SOLAR RADIATION AND THE WHOLE PLANET BECAME WAY TOO HOT, ENOUGH TO MELT LEAD METAL.

EARTH IS THE NEXT HOTHOUSE IN OUR SOLAR SYSTEM, FOR THE SAME REASON, UNLESS WE DECLARE A WORLDWIDE EMERGENCY FUND FOR **GIGATON CO2 REMOVAL** TO STOP AN EXTINCTION EVENT ONLY ONE CENTURY AWAY.

TABLE OF CONTENTS

Introduction	10
Chapter 1: Overview of Real Global Warming Cause and History	13
Chapter 2: Understanding the Hansen Equation	24
Chapter 3: The Gigaton Challenge	36
Chapter 4: Role of Billionaires and Trillionaires in Climate Action	49
Chapter 5: Innovative Technologies for Carbon Capture	59
Chapter 6: Community-Driven Initiatives and Success Stories	70
Chapter 7: Interdisciplinary Approaches to Climate Solutions	79
Chapter 8: Practical Steps for Individuals	90
Chapter 9: Global Perspectives and Cooperation	100
Chapter 10: Addressing Common Objections	110
Chapter 11: Education and Advocacy for Systemic Change	121
Chapter 12: Evaluating and Improving Carbon Removal Strategies	130
Conclusion	140
References	143

Note: Most of the slides used in this book are available online in the video slideshow lecture by Dr. Valone at https://tinyurl.com/ClimateVideoValone.

Introduction

In 2006, a groundbreaking article appeared in NASA's "Technology Review," penned by the esteemed climatologist James Hansen. This seminal work laid bare the stark reality of our planet's climate crisis, revealing the precise, linear relationship between Earth's temperature, CO2 levels, and sea level rise. Hansen's meticulous analysis of the Vostok Ice Core data, spanning an astonishing 420,000 years, provided irrefutable evidence of the intricate dance between these critical variables.

Shockingly, this pivotal research seems to have escaped the attention of the vast majority of the world's scientific community. Yet, within its pages lies a beacon of hope – a path to reversing the dire consequences of our climate emergency. Hansen's work not only illuminated the problem but also hinted at the solution, demonstrating the potential to undo the damage wrought by our reckless experimentation with Earth's delicate balance.

In this book, I invite you to embark on a journey of discovery as we explore the profound implications of what I have termed the "Hansen Equation." Together, we will unravel the complexities of this elegant formula, derived from Hansen's painstaking analysis, and reveal its power to guide us toward a more sustainable future.

For those among you who seek the truth through scientific evidence, I offer a glimmer of optimism amidst the gloom. By following the roadmap laid out in these pages, we can chart a course back to the more temperate climes of the 1950s, when atmospheric CO2 levels were in harmony with our planet's natural rhythms. It won't be easy, but the alternative – a world ravaged by the relentless march of hothouse climate change – is simply unthinkable.

As we delve deeper into the science behind the greenhouse effect, you will come to understand the immense heat-trapping potential of CO2 and the devastating consequences of our actions. For the past million years, Mother Earth has maintained a delicate equilibrium, with CO2 levels fluctuating within a narrow range, **always BELOW 290 ppm**. Yet, in the blink of an eye, humans have recklessly pushed these levels 40% higher, unleashing a cascade of catastrophic effects that threaten our very existence. The best proof is what Hawaii tells us:

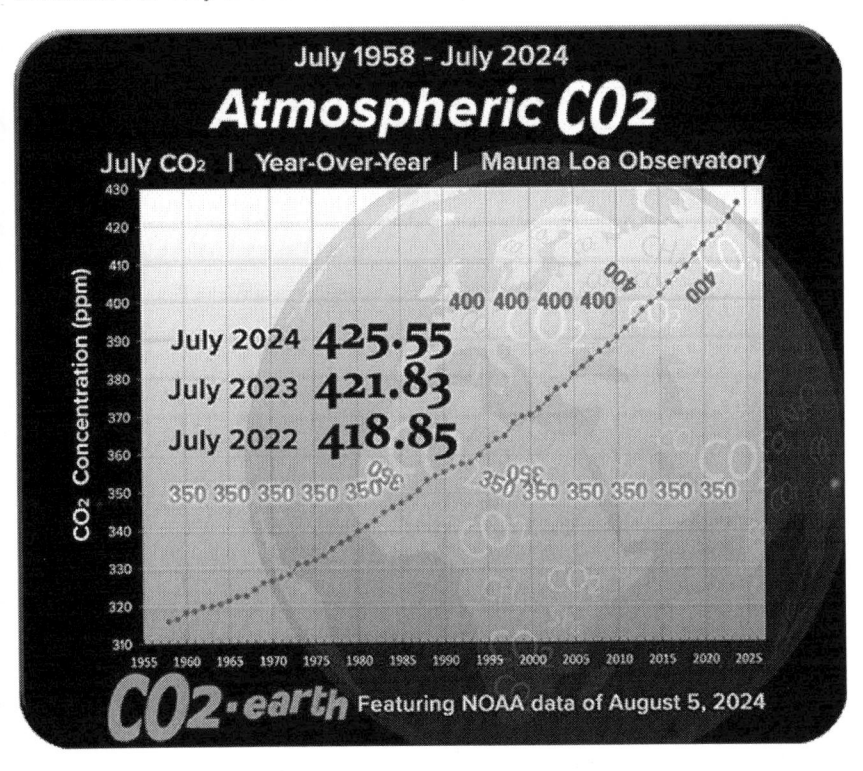

July 1958 - July 2024
Atmospheric CO_2
July CO_2 | Year-Over-Year | Mauna Loa Observatory

July 2024 **425.55**
July 2023 **421.83**
July 2022 **418.85**

CO_2·earth Featuring NOAA data of August 5, 2024

The above graph, thanks to https://www.co2.earth/ and the **National Oceanic and Atmospheric Administration** (NOAA), shows the increasingly dense CO2 with no signs of slowing down. Not only are the skeptics deceiving the public into complacency ("We still have time" – Elon Musk, 2024) but to think the global temperature will level off soon by any quick *renewable means* when even the CO2 rate is going up *faster and faster* is sheer fantasy. This book conveys the best truthful condition of the atmosphere that you need to know, from the world's expert climatologists. The same CO2.earth website, on the next page, also exposes the true **"CO2 Emissions RATE"** in ppm/year which all of us hoped would be constant, or even starting to decrease slightly so the CO2 emissions level per year would peak.

Decade	(ppm per year)
2011 - 2020	2.43
2001 - 2010	2.04
1991 - 2000	1.55
1981 - 1990	1.56
1971 - 1980	1.35
1961 - 1970	0.91

ppm = parts per million

This chart shows <u>extra</u> amounts of CO2 that we have ADDED each year *to the air*. **NOTE:** the RATE in ppm/yr is also increasing almost every decade and possibly becoming *exponential*. Compare with the Carbon Emissions per Annum in the next chapter. Therefore, we should begin now to **remove gigatons of heat-trapping CO2 from our atmosphere**, a Herculean task that will require the collective efforts of visionary leaders and innovative companies alike. Thankfully, many have already risen to the challenge, spearheading ambitious carbon dioxide removal (CDR) initiatives to turn the tide.

In the pages that follow, we will celebrate these unsung heroes, the vanguard in the battle to save our planet for generations to come. Their tireless work and unwavering commitment serve as an inspiration to us all, a clarion call to action in the face of an existential threat. The inserted slides in this book are from my IEEE (ISTAS) presentation at Tufts University a few years ago which I am happy to share with you.

As a physicist, engineer, and inventor with over three decades of experience, I have dedicated my career to exploring the frontiers of science and technology. From cutting-edge electrotherapy devices to groundbreaking research in zero-point energy, I have sought to push the boundaries of what is possible. Now, I turn my attention to the most pressing challenge of our time, the coming hothouse earth, armed with the knowledge that solutions exist if only we have the courage to embrace them.

Join me on this transformative journey as we confront the stark realities of our climate crisis head-on and work together with only a few decades left to forge a brighter, more sustainable future for all. The road ahead may be difficult, but with determination, ingenuity, precisely accurate facts about our planet's history, and the power of science on our side, we can and will prevail.

Chapter 1

Overview of the Real Global Warming Cause and History

In the world of climate science, few moments have sparked as much intrigue and debate as the pivotal revelations by James Hansen, published by MIT in 2006 (*Technology Review*, July/August 2006). The significance of his findings, shown here in a closeup, segmented version of his "Hansen graph," cannot be overstated. Obtained from ancient ice, it is an accurate history of Earth's atmospheric past. In this intricate design lies the Vostok Ice Core data, a remarkable collection of evidence spanning 420,000 years, revealing the linear relationship between Earth's temperature, CO2 levels, and sea levels. Hansen's Graph is a meticulous work **proving how temperature tracks the increases in CO2 and vice versa.** This wasn't just an academic exercise; it was a clarion call, a startling exposé that laid bare the very mechanics of our planet's climate system, with a clear predictive nature for the future climate expectations. He also published in www.pnas.org_cgi_doi_10.1073_pnas.0606291103 during 2006, where his highest global temperature change projection for 2020 was for 1.5°C which came true (at 0.35°C/decade back then). Today, we are adding at least 2 ppm CO2 EVERY YEAR! Accumulating another 20 ppm for an additional degree of indebted temperature rise is now a **decade event.** *Stop reading* and go to https://tinyurl.com/GlobalTempCO2 to see a 1-minute CO2 video.

Hansen's graphed analysis was not merely about observing trends; it demonstrated the profound interconnectedness of ALL of the climate variables. It was a revelation that should have echoed through the halls of science, yet it somehow slipped under the radar of many experts. This wasn't just a missed opportunity; it was a critical moment that, if heeded, could have shifted the course of climate discourse. Now, about 20 years later, Dr. Hansen's work showed us that the changes we are witnessing today are not anomalies but part of a larger, predictable pattern. His insights were not only about *understanding the past accurately* but about providing a roadmap for the future, a future where we have the power to reverse the damage if we act boldly, decisively, *and with vigor,* echoing President John F. Kennedy.

1.1 Climate Temperature, CO2 Rise, and Sea Level Expectations

In 2006, a significant moment in climate science unfolded when James Hansen revealed in his graph below, a linear and very tight correlation between Earth's temperature and CO2 levels. The discovery was based on Vostok Ice Core data, a comprehensive record spanning 420,000 years, meticulously documenting the close interplay between temperature, carbon dioxide, and sea levels.

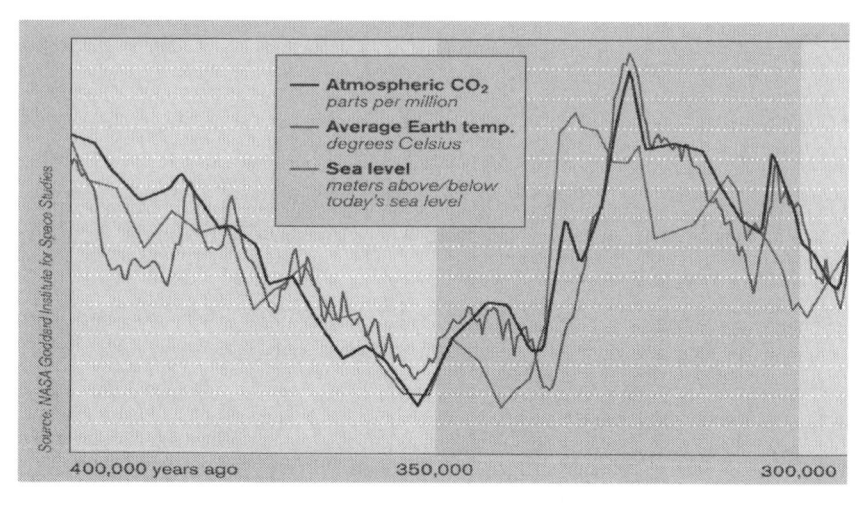

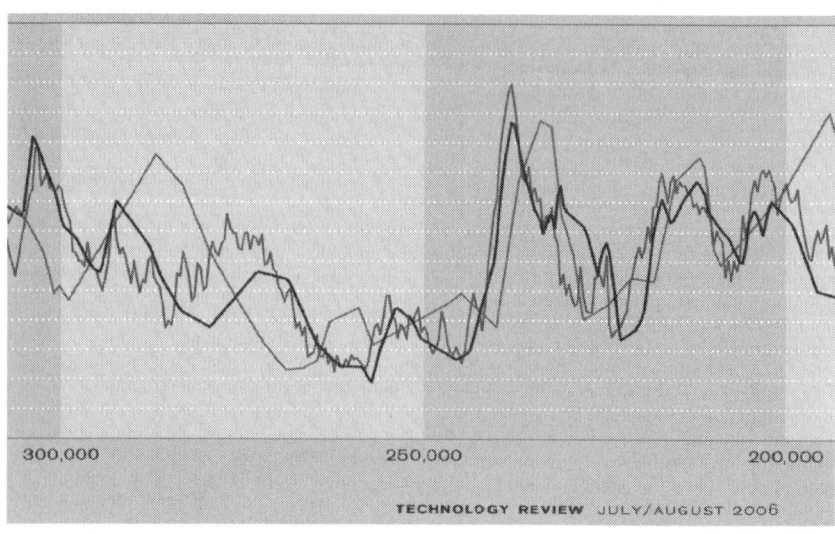

— Atmospheric CO_2
parts per million

— Average Earth temp.
degrees Celsius

— Sea level
meters above/below
today's sea level

Source: NASA Goddard Institute for Space Studies

400,000 years ago — 350,000 — 300,000 — 250,000 — 200,000

TECHNOLOGY REVIEW JULY/AUGUST 2006

14

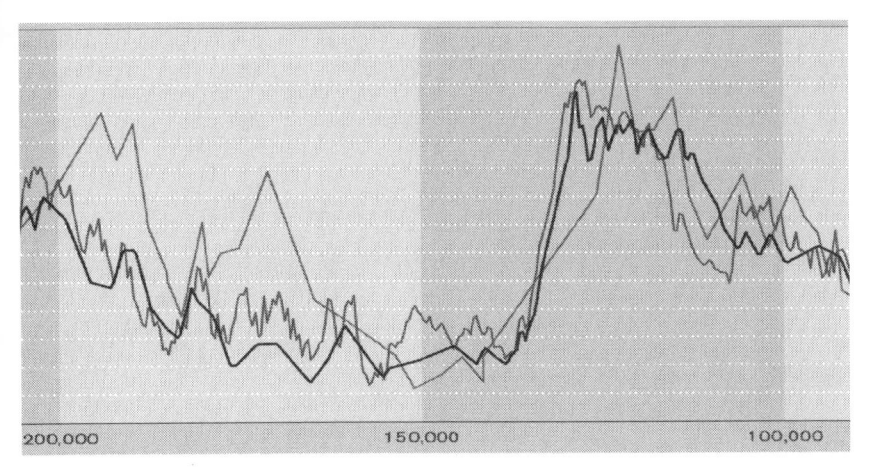

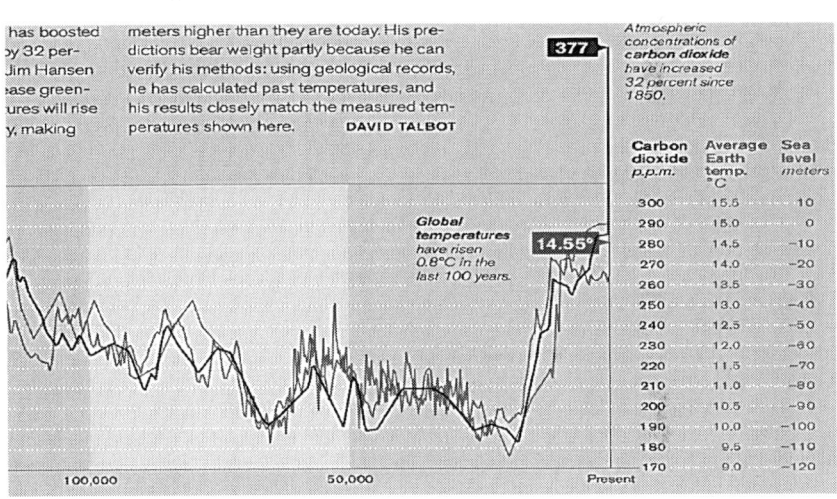

has boosted	meters higher than they are today. His pre-		Atmospheric concentrations of carbon dioxide have increased 32 percent since 1850.			
by 32 per-	dictions bear weight partly because he can	377				
Jim Hansen	verify his methods: using geological records,					
ase green-	he has calculated past temperatures, and					
ures will rise	his results closely match the measured tem-					
y, making	peratures shown here. DAVID TALBOT					

Carbon dioxide p.p.m.	Average Earth temp. °C	Sea level meters
300	15.5	10
290	15.0	0
280	14.5	−10
270	14.0	−20
260	13.5	−30
250	13.0	−40
240	12.5	−50
230	12.0	−60
220	11.5	−70
210	11.0	−80
200	10.5	−90
190	10.0	−100
180	9.5	−110
170	9.0	−120

Global temperatures have risen 0.8°C in the last 100 years. **14.55**

This revelation was not just another scientific finding; it was a pivotal moment that highlighted the profound impact of human activities on our planet's climate system. **Note:** to see the **full-color** 2006 Hansen graph altogether in the same detail but as one single graph, including the data table he assembled for it, reprinted on our IRI website link: https://integrityresearchinstitute.org/climatechart.pdf. Also notice the CO2 level in 2006 was at 377 ppm vs. 420 ppm in 2025. For the BW printing, it may help the reader to note that at 400,000 years ago, the black top line is CO2, the next blue line below it is sea level, and the lowest red line is the average earth temperature. It is very important to note that each of them keep alternating in taking the lead as time goes on, thus proving the true, equally weighted correlation the three variables exhibit here. The Hansen graph also showcases the **four (4) previous ice ages** (minimum levels) very clearly.

For centuries, Earth's climate has been shaped by natural processes, with cycles of warming and cooling driven by factors such as volcanic eruptions, solar radiation, and ocean currents. However, the rapid increase in CO2 levels since the Industrial Revolution has disrupted this delicate balance. Human activities, primarily the burning of fossil fuels, have released vast amounts of carbon dioxide into the atmosphere, leading to a significant rise in global temperatures. This unprecedented increase in CO2 levels has far-reaching consequences, including rising sea levels, extreme weather events, and disruptions to ecosystems worldwide. The fact that **CO2 levels are expected to keep increasing**, instead of decreasing as IPCC would hope for, is shown graphically in the World Energy Usage that is 90% fossil-fueled technology and in the "unmitigated growth of carbon emissions" slide.

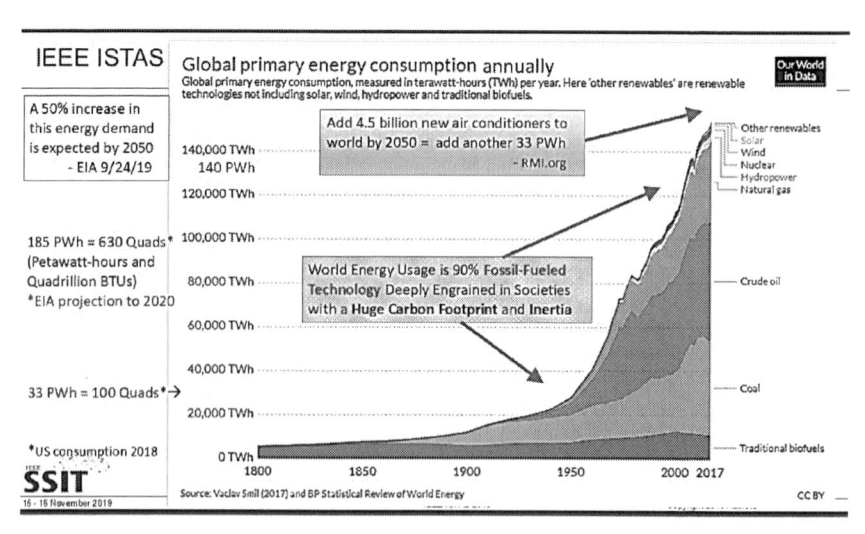

Hansen's work demonstrates that the relationship between temperature and CO2 is surprisingly very linear. Therefore, it is not merely coincidental but a fundamental aspect of Earth's climate system: global temperature change is really proportional to the heat-trapping CO2 average levels, *though one may exceed the other for a while before the other two catch up again.* His findings provided the scientific community with a critical understanding of how human-induced CO2 emissions are driving global warming and its associated impacts. The data from the Vostok Ice Core offered a unique perspective on the historical climate, revealing the sources that have shaped our planet's climate over millennia, never exceeding 290 ppm of the heat-trapping gas CO2 for mega-years. **Carbon Emissions per Annum** (Global Carbon Budget, essd.copernicus.org) below show the excessive world's CO2 exhaust in billions of tons (gigatons) per year.

IEEE ISTAS 2019 **CARBON EMISSIONS** PER ANNUM

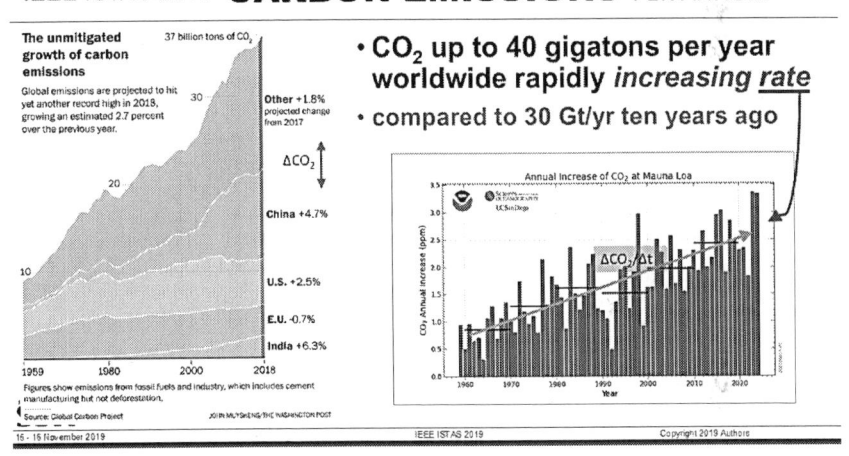

The unmitigated growth of carbon emissions
Global emissions are projected to hit yet another record high in 2018, growing an estimated 2.7 percent over the previous year.

37 billion tons of CO₂
Other +1.8% projected change from 2017
ΔCO₂
China +4.7%
U.S. +2.5%
E.U. -0.7%
India +6.3%

Figures show emissions from fossil fuels and industry, which includes cement manufacturing but not deforestation.
Source: Global Carbon Project
JOHN MUYSKENS/THE WASHINGTON POST
15 - 15 November 2019

- **CO₂ up to 40 gigatons per year worldwide rapidly *increasing* rate**
- **compared to 30 Gt/yr ten years ago**

Annual Increase of CO₂ at Mauna Loa
ΔCO₂/Δt

IEEE ISTAS 2019   Copyright 2019 Authors

The implications of Hansen's research are profound. History doesn't lie. It is destined to repeat itself too. By establishing a clear causal link between CO2 levels and temperature, his work underscored the urgent need for action to mitigate the effects of climate change. It highlighted the importance of reducing carbon emissions and transitioning to renewable energy sources. Hansen's findings serve as a stark reminder that history will repeat itself given the same driving conditions and therefore, the choices we make today will determine the <u>very predictable hothouse future</u> of our planet.

The linear correlation between temperature and CO_2 levels is a powerful tool for understanding the past and also for reliably predicting future climate change, especially in times when CO_2 may temporarily exceed the corresponding temperature average worldwide. However, the increase in CO_2 alone seems to be **exponentially increasing**, from the latest direct atmospheric CO_2 measurements from the Mauna Loa and South Pole stations.

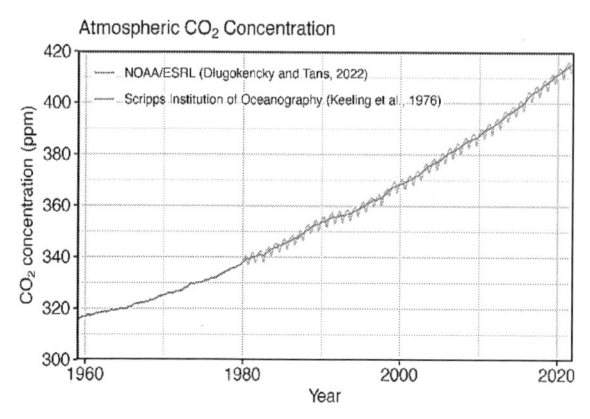

What humans are doing to the climate today has historically already occurred on earth about 56 million years ago during the **Paleo-Eocene Thermal Maximum (PETM)** as seen in the Paleoclimatology graph below, matching the CO_2 levels in parts per million then and what we expect by the end of THIS century.

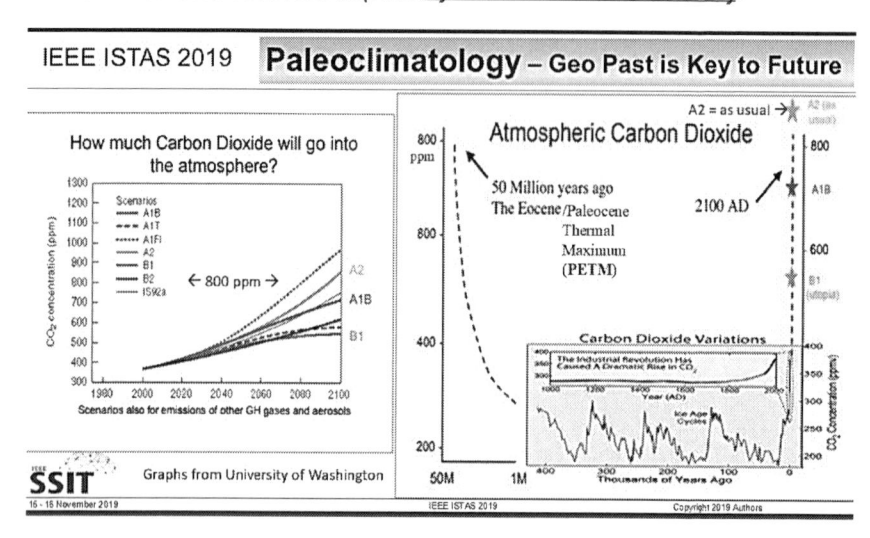

Notice that the **Atmospheric CO2 Concentration** graph will most likely surpass 520 ppm in another sixty years (by 2080) as it increases. If it ever becomes linearly increasing, *at that point it can be determined that the RATE has peaked.* This is explained at the end of Chapter 3 with the World Population graph, where the quantity peaks about 100 years later! **Note:** The +100 ppm = +5°C and is added to the +6°C we are ALREADY indebted for (using the Hansen Eq. in Chapter 2). This puts us past +11°C by 2100 and brings to mind the PETM.

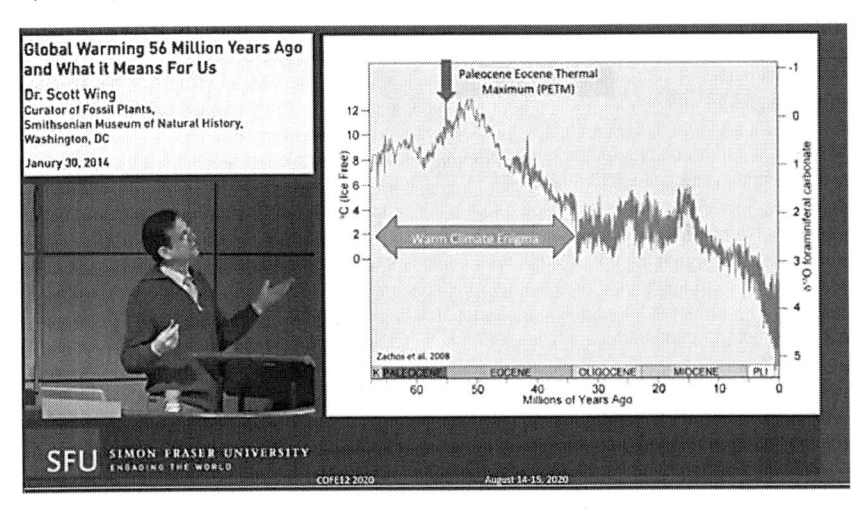

At this point, we must take a turn into the **PETM** briefly to give the reader a view of what our Dr. Scott Wing here in DC at the Natural History Museum says about our future. It was only a couple of million years after the meteor strike 60 MY ago wiped out the dinosaurs that the historic PETM occurred with up to +12°C higher than our same baseline of about 15°C that climatologists use for a "zero" point.

In the following sections, we will delve deeper into the science behind climate change, exploring the mechanisms that drive global warming, the role of carbon dioxide as a greenhouse gas, and the potential solutions for mitigating its effects. Through a comprehensive analysis of the latest research and data, we aim to provide a nuanced understanding of the challenges and opportunities that lie ahead in the fight against climate change.

IEEE ISTAS 2019   ALASKA Columbia Glacier ONLY SIX Years Apart

Columbia Bay, Alaska – Photographer James Balog, Nat. Geo. magazine: **Extreme Ice Survey of 18 Glaciers**
The most extreme: Columbia Glacier is losing one mile every three years – so two miles of loss are shown below.
Since 1980, this glacier has lost height equal to the Empire State Building!

2006   2012

The Vostok Ice Core data serves as a crucial piece of the puzzle, offering invaluable insights into the historical climate and the factors that have influenced it over time. By examining this data in conjunction with modern climate models and observations, we can gain a more comprehensive understanding of the processes that govern Earth's climate and the role that human activities play in shaping its future.

As we explore the intricacies of climate change, it is important to approach the subject with an open mind and a willingness to learn. The science of climate change is complex and constantly evolving, with new discoveries and advancements being made every day. By staying informed and engaged, we can contribute to the ongoing dialogue and work towards solutions that address the root causes of climate change while promoting sustainability and resilience.

In the coming chapters, we will explore the various factors that contribute to climate change, the potential impacts of rising temperatures and CO2 levels, and the strategies for reducing carbon emissions and mitigating their effects. Through a combination of scientific analysis, practical solutions, and real-world examples, we aim to empower readers with the knowledge and tools needed to make informed decisions and take meaningful action in the fight against climate change.

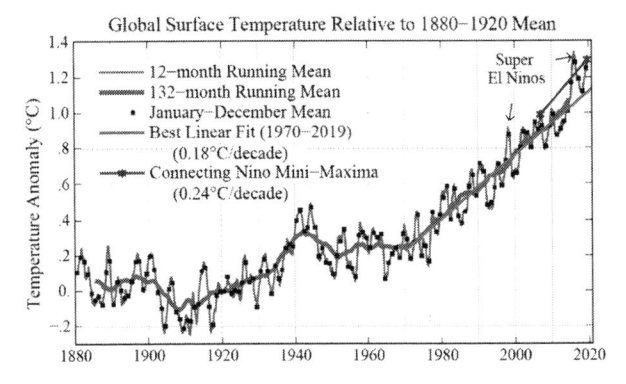

Global Surface Temperature Relative to 1880–1920 Mean

[Graph: Temperature Anomaly (°C) vs year 1880–2020, showing 12-month Running Mean, 132-month Running Mean, January–December Mean, Best Linear Fit (1970–2019) (0.18°C/decade), Connecting Nino Mini–Maxima (0.24°C/decade), with Super El Ninos marked]

The Temperature graph above is credited to NOAA, at which point we must point out the highest rate they report is **only 0.24°C/decade**. Compare this number with Dr. Hansen's higher measured rate starting from 2010 onward, as seen from *the same graph* in Chapter 2.

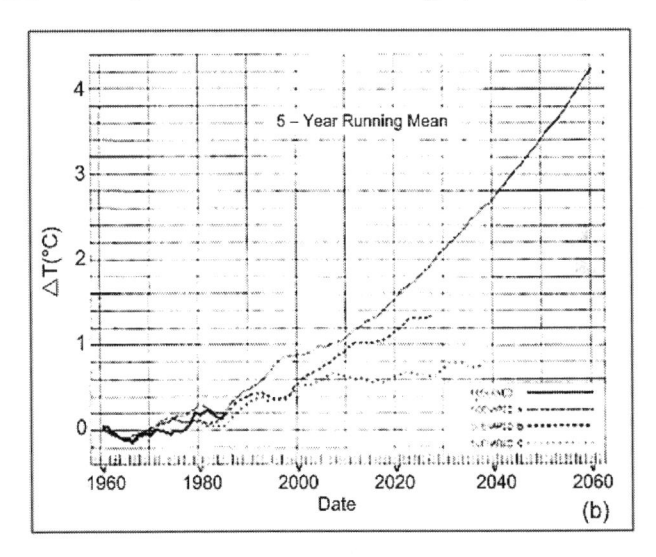

[Graph: ΔT(°C) vs Date 1960–2060, 5-Year Running Mean (b)]

Above is **the original Hansen temperature projection to 2060** that he published in 1988 which has been realized almost exactly as time has gone on (Hansen, 1988). The projection from now to 2060 will also be fulfilled in most likelihood since there is no major effort at CDR yet.

This puts us at about a +6°C realization by 2100 approximately with an indebtedness *past* the **PETM of +12°C into the 22nd century**, which if +6°C is not regarded as an extinction event threshold, then the +12°C and beyond certainly will be a real extinction event.

Hansen's work serves as a call to action, urging us to take responsibility for our actions and make informed decisions that prioritize the health and well-being of our planet. By embracing the findings of his pivotal research, we can chart a course toward a more sustainable and equitable future for all.

His discoveries tell us what science says about our future and not this author's personal bias. If it seems pessimistic, think twice about the facts presented and where you think there is any possible error. It took me twenty years for this book to emerge in 2026 since Dr. Hansen shocked everyone about our planet's historical connectedness between CO2, Temperature, and sea level in 2006 and suffered great derision at Congressional Hearings in the 1980's as a result. However, time has proven his NASA computer projections to be the most prophetically accurate.

The path ahead may be challenging, but it is also filled with opportunities for innovation, collaboration, and positive change. By coming together as a global community, we can address the pressing issues of climate change and work towards a more sustainable and equitable future for all. Whether you are a student, an environmentalist, a policymaker, or simply someone who cares about the future of our planet, your voice and actions matter. Together, we can make a difference and create a world that is not only resilient to the impacts of climate change but also thriving in harmony with the natural environment.

Reflection Section

As we navigate the complexities of climate change from the previous graphs (essd.copernicus.org), it is essential to recognize the interconnectedness of Earth's climate system and the role that human activities play in shaping its trajectory. For those who still may have questions about the impact of CO2 on heat absorption and atmospheric temperature increase, notice the similarities between the Atmospheric CO2 Concentration and the Global Surface Temperature graphs in this chapter. Both are going up and if the recent section from 1960 or even better, from 2010 in the Global Surface Temperature graph was shown, as Brown and Calderia from Stanford University do in Chapter 3, we would see the correlation between the CO2 trend and the temperature trend. Please view the most dramatic classroom **"Greenhouse Gas Demo"** done by a school teacher on YouTube in 4 minutes that shows a CO2 experiment with heat lamps: https://tinyurl.com/CO2heatdemo.

Each **Reflection Section** should give the reader pause to think about how serious the earth's situation is, how confusing the media reporting is regarding climate, what the facts and especially the future trends really are, and especially, what DAC and CDR solutions can be amplified to read the gigaton level to actually make a dent in the heat blanket above our heads that keeps getting thicker each year as all of the world's glaciers keep getting thinner each year. Reflect on what a world without a heat sink will be like as the poles become ice-free soon. A second homework assignment should be to watch the free NatGeo documentary on YouTube **"Six Degrees Could Change the World"**. https://youtu.be/EU5tUY3W3WI. They are talking degrees C as well. See if you can survive watching the last half of the video without squirming as the final +5°C and +6°C are portrayed with graphic detail. It is still the most accurate projection of what the last half of our current century will endure as the ones who can afford them try to purchase billions of air conditioners to avoid the heat. Stanford University agrees, as you will see in Chapter 3.

Chapter 2

Understanding the Hansen Equation

Imagine venturing into the depths of Antarctica, where the ice holds untold secrets and stories yet to be discovered. Here lies the Vostok Station, an isolated research base perched upon one of the thickest ice sheets in the world. Beneath its frozen surface, the secrets of Earth's climate history are preserved in the Vostok Ice Core. These cores are more than just frozen columns; they are time capsules that offer a glimpse into our planet's past, revealing the intricate dance of climate change over hundreds of thousands of years. For over four kilometers beneath the surface, these ice cores have remained untouched, safeguarding the atmospheric records of epochs gone by. As scientists carefully extract these cores, they reveal layers upon layers of ice, each one a snapshot of a moment in time, offering rich insights into the conditions that shaped our climate.

The extraction process itself is a marvel of precision and technology. Imagine a delicate operation where every fragment of ice is meticulously preserved, ensuring that the trapped air bubbles remain intact. These bubbles are not mere pockets of air; they are tiny windows into the past, holding within them the composition of ancient atmospheres. By analyzing these bubbles, scientists can determine the levels of carbon dioxide, methane, and other gases that existed at different points in history. This data (see Chapter 1) is invaluable, providing a detailed record of how these gases have fluctuated over millennia, which yields the simple **Hansen Equation** in this chapter. The information gleaned from these cores is not just about gases; it also includes temperature proxies, which help us understand the climate conditions that prevailed in these ancient times.

The Vostok Ice Core data spans an impressive 420,000 years, covering multiple glacial and interglacial periods. These cycles, marked by the advance and retreat of ice sheets, offer a unique perspective on the natural rhythms of Earth's climate. During glacial periods, lower temperatures and higher ice volumes prevail, while interglacial periods bring warmer climates and receding ice. By examining the core data, scientists can trace these transitions, observing how carbon dioxide concentrations and temperatures rise and fall in tandem. This extended timeline not only enhances our understanding of past climate dynamics but also serves as a crucial reference for modern climate models. The richness of this data allows for rigorous testing and validation of these models, ensuring that our predictions are grounded in historical reality.

The Vostok Ice Core revelations have become foundational to the science of climate modeling. By linking past climate data with current models, scientists can refine their predictions and enhance the accuracy of future climate scenarios. This is not a matter of guesswork; it's a meticulous process of aligning historical records with contemporary observations, creating a comprehensive picture of Earth's climate system. The Vostok data, with its unparalleled depth and detail, stands alongside other ice core records, forming a robust framework for understanding the complexities of global climate change. As we integrate this information, we gain the ability to simulate future climates, providing a critical tool for policymakers and researchers alike.

2.1 Interactive Element: Reflecting on the Past

The Vostok Ice Core revelations offer more than just scientific insights; they provide a narrative that connects us to the Earth's history. They remind us that our planet has undergone significant changes long before human influence, yet our current trajectory is unprecedented in speed and scale. As you delve into these revelations, consider the broader implications for our understanding of climate change. The data not only tells a story of past climates but also challenges us to confront the reality of our impact on the planet. Through this exploration, we gain a deeper appreciation for the delicate balance of Earth's systems and the urgent need to restore harmony in the face of an uncertain future.

2.2 Linear Correlations: CO2, Temperature, and Sea Levels

James Hansen's research opened the door to understanding how intricately linked CO2 levels, temperature changes, and sea levels are. By meticulously plotting data points over hundreds of thousands of years, Hansen revealed a clear, linear relationship that connects these three critical components of our climate system (in the Hansen graph from previous chapter). Imagine a graph where each axis represents one of these variables: as CO2 concentrations increase, temperatures rise, and subsequently, sea levels follow suit. This visualization is not just abstract science—it is a tangible demonstration of the impact of heat-trapping carbon dioxide.

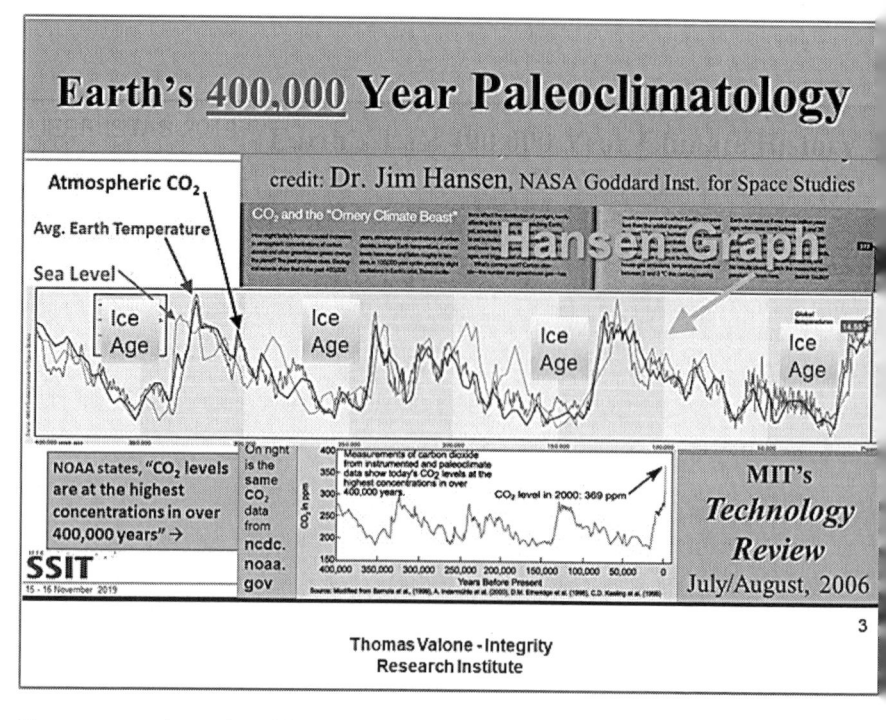

To summarize the Hansen graph's importance, viewing it here comprehensively, it is easy to see how close the three variables track each other over four ice ages. Humans have created what Hansen calls the "Ornery Climate Beast" by thermally forcing Mother Nature above natural limits, not preparing for her formidable consequences.

Through quantitative analysis, Hansen's work underscores that even small changes in atmospheric CO_2 can lead to significant linear correlated shifts in global temperatures, which in turn drive sea level rise. The big surprise to this author and most every climatologist is that this relationship is pivotal and *entirely linear*. Therefore, it provides a reliable and easy-to-calculate framework for predicting future climate scenarios, helping us understand the trajectory we're currently on. It must be emphasized that all leading climatologists rely upon computer-generated climate modeling programs.

In this book, instead of relying on "climate models" churned out by computers with a limited number of known variables which invariably never include all that Mother Nature throws at us, *earth's history does not lie*. Moreover, the Hansen graph of earth's complete history includes *ALL of the missing variables,* including those that interact and multiply effects that computer models always forget. Therefore, the Hansen graph must be the most reliable, honest, and comprehensive by necessity. For example, the release of methane from melting permafrost is already included in the historical record if the same conditions were present in previous ages.

The implications of these correlations are profound, offering valuable insights into the dynamics of climate change and the potential paths it might take. Predictive models, built upon these linear relationships, allow scientists to simulate future climate conditions, taking into account various emissions scenarios. These models predict that if current trends continue, we could see unprecedented temperature increases and sea level rises within this century. Feedback mechanisms, such as the ice-albedo effect, where melting ice reduces Earth's reflectivity and accelerates warming, further complicate these predictions. Understanding this feedback is critical, as it can amplify the effects of initial changes, leading to cascading impacts throughout the climate system. Armed with this knowledge, we can better plan and implement strategies to mitigate these effects by either reducing emissions or developing adaptive measures to protect vulnerable communities.

The scientific community widely acknowledges the correlations identified by Hansen, with numerous peer-reviewed studies and international climate reports reinforcing these findings. The **Intergovernmental Panel on Climate Change** (IPCC) and other authoritative bodies have long supported the idea that CO2, temperature, and sea levels are inextricably linked and that Carbon Dioxide Removal (CDR) is also a necessary part of remediation. This consensus is crucial, as it forms the basis for global climate policy and action. By grounding our understanding in robust scientific evidence, we can confidently advocate for measures to curb emissions and limit further damage to our environment. The agreement across the scientific community serves as a powerful tool in convincing policymakers and the public of the urgent need for action.

Despite the strong linear correlations, anomalies and challenges persist in the data, reminding us that the Earth's climate system is complex and influenced by a myriad of factors. External variables, such as volcanic eruptions or solar variations, can temporarily alter climate patterns, leading to outliers in the data. Technological limitations in past data collection also contribute to uncertainties. For instance, early methods of measuring gas concentrations were less precise than today's advanced techniques. These challenges highlight the importance of continuous research and improvements in climate science. By acknowledging these anomalies, we can refine our models and predictions, ensuring that they accurately reflect the multifaceted nature of our planet's climate.

2.3 Unveiling the Reversibility of the Hansen Equation

The concept of climate reversibility might seem like a distant hope, but it is rooted in solid scientific principles. At its core, reversibility suggests that if we can lower atmospheric CO2 levels, we might also reverse the warming trends and halt rising sea levels. This idea isn't just theoretical; history provides examples, albeit on slower, natural timescales. After volcanic eruptions, CO2 levels have sometimes dropped, leading to temporary cooling. Such instances highlight the possibility that human intervention could mimic these natural processes. The Hansen Equation, derived from James Hansen's insights, offers a framework to understand and potentially guide this reversal. It suggests that by reducing CO2, we can trigger a domino effect—temperature and sea level changes will follow suit, albeit with some lag due to the ocean's massive heat capacity.

The Hansen Equation below isn't just a theoretical construct; it is a tool *extracted from the most precise historical graph of our planet's past 420,000 years* that can guide real-world climate action, if we learn our lesson from the past. It is a mathematical formula that describes the relationship between CO2 levels, temperatures, and sea levels. The last two have a time delay after CO2 changes, with sea level being the slowest to respond but **the end result is quantitative** with the Hansen Equation! By inputting current CO2 concentrations, the equation can help predict corresponding temperature changes and sea level shifts. *This predictability should be crucial for policymakers,* providing a scientific basis for setting carbon reduction targets. If we know the likely outcomes of reducing CO2 by a certain

$$+/- (20 \text{ ppm CO2} = 1 °C = 20 \text{ m sea rise})$$

amount, we can tailor policies to achieve those goals. It's akin to having a map that not only shows where we are but where we will be and also, where we could be if we take the right steps.

Achieving the reversibility envisioned by this **Hansen Equation** requires a multifaceted approach. One of the most promising strategies involves carbon capture technologies, which can remove CO2 directly from the atmosphere. These technologies, ranging from large-scale industrial solutions to smaller, localized systems, promise to make a significant dent in atmospheric carbon levels. Alongside this, integrating renewable energy sources like wind and solar into our power grids can reduce the amount of new CO2 being emitted. By shifting away from fossil fuels, we cut off the tap, so to speak. Reforestation and changes in land use also play a crucial role. Trees naturally capture carbon, and by restoring forests or adopting more sustainable agricultural practices, we can enhance this natural process. Together, these strategies form a cohesive plan not only to stop further damage but also to begin the healing process.

Despite the promise of this **Hansen equation's +/- reversibility**, several challenges stand in the way of implementation. Economic barriers are significant; the cost of deploying carbon capture technologies on a global scale is immense. Many developing nations, already struggling with economic challenges, may find it difficult to invest in such solutions without international support. Political resistance also poses a formidable obstacle. Climate change policies often face pushback from industries reliant on fossil fuels and even from political leaders skeptical of the science. Overcoming these challenges requires not just technological innovation but a concerted effort to build political will and financial mechanisms that support climate action. It means fostering collaborations between nations, industries, and communities to share resources and knowledge, creating a unified front against climate change.

Amazingly, the REVERSIBILITY of the Hansen graph shows that if and when we bring the CO2 levels back down to 300 ppm or below, the polar ice sheets will start forming again, albeit very slowly, but the average global temperature will start to come back down even faster in only a few decades.

2.4 Historical Perspectives on Climate and CO2 Levels

Throughout Earth's history, the climate has experienced periods of significant change, with CO2 levels playing a crucial role in these transformations. Before the Industrial Revolution, CO2 concentrations remained relatively stable, fluctuating slightly during glacial and interglacial periods. These changes were driven by natural processes, such as volcanic activity and oceanic cycles, which influenced atmospheric composition over millennia. The pre-industrial era, however, was a time when the balance between carbon emissions and absorption was maintained naturally. For thousands of years, human activities had minimal impact on this delicate balance. It wasn't until the dawn of the Industrial Revolution that we began to see a dramatic shift. This period marked a turning point in the Earth's atmospheric history. With the advent of steam engines and the widespread burning of coal, carbon emissions skyrocketed, setting in motion a chain of events that would have lasting consequences on our planet's climate.

As we examine historical CO2 levels and their correlation with temperature changes, we can draw parallels between past and present climate conditions. During ice age cycles, lower CO2 concentrations were always associated with cooler global temperatures as seen in the Hansen graph. These periods, characterized by extensive ice sheets covering large parts of the Earth's surface, serve as stark reminders of the power of carbon dioxide in shaping climate patterns. Conversely, interglacial periods with higher CO2 levels were marked by warmer temperatures and retreating ice. This cyclical pattern of warming and cooling, influenced by CO2 fluctuations, provides valuable insights into the natural variability of Earth's climate system. Today's rising CO2 levels, past the traditional 290 ppm maximum, mirror those of past interglacial periods, but the current level of CO2 as well as its rate of increase (see "unmitigated growth" graph) is unprecedented. This rapid change underscores the significant impact of human activity on the climate, highlighting the urgent need to address the root causes of global warming.

"Those who do not learn from history are bound to repeat it"

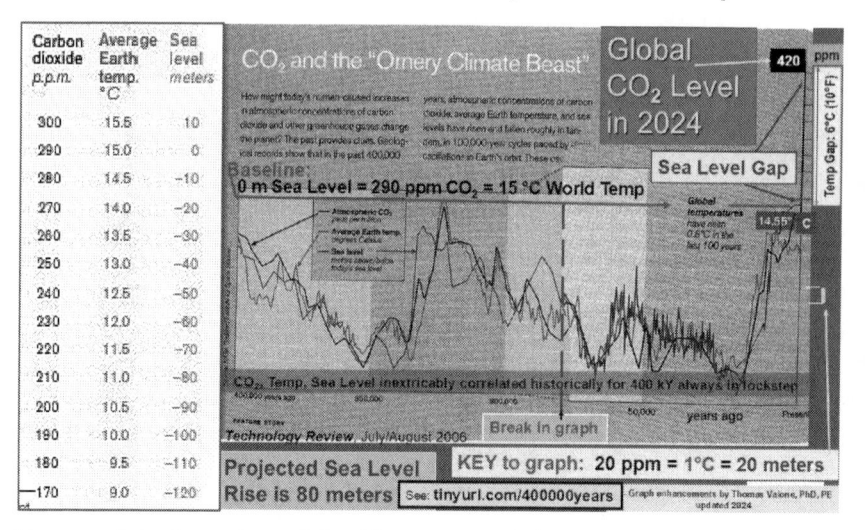

Carbon dioxide p.p.m.	Average Earth temp. °C	Sea level meters
300	15.5	10
290	15.0	0
280	14.5	-10
270	14.0	-20
260	13.5	-30
250	13.0	-40
240	12.5	-50
230	12.0	-60
220	11.5	-70
210	11.0	-80
200	10.5	-90
190	10.0	-100
180	9.5	-110
170	9.0	-120

CO$_2$ and the "Ornery Climate Beast"

Global CO$_2$ Level in 2024 — 420 ppm

Baseline: 0 m Sea Level = 290 ppm CO$_2$ = 15 °C World Temp

Sea Level Gap

CO$_2$, Temp, Sea Level inextricably correlated historically for 400 kY always in lockstep

Technology Review, July/August 2006 — Break in graph

Projected Sea Level Rise is 80 meters. See: tinyurl.com/400000years

KEY to graph: 20 ppm = 1°C = 20 meters

Graph enhancements by Thomas Valone, PhD, PE updated 2024

My annotated version of the Hansen graph is shown here and online at https://tinyurl.com/SixDegreeGraph, truncated in the middle, was updated in 2024 when the CO2 levels reached 420 ppm. It helps the reader see how simple the calculation is to find the indebtedness for the guaranteed average atmospheric temperature rise which was +6°C at that time with about +2°C already realized in 2025. It will take a few hundred years for the sea level gap to close but there is enough landlocked ice in melting glaciers to almost reach an astounding 70 meters at least. Remember that 2/3 of the earth's glaciers will be gone by the end of this century.

In understanding climate science, we owe much to the pioneering climatologists whose groundbreaking research has shaped our current knowledge. Figures like Charles Keeling, who meticulously documented the rise of atmospheric CO2 with the Keeling Curve (see Reference section), provided compelling evidence of the ongoing increase in greenhouse gases. Likewise, Svante Arrhenius's early work on the greenhouse effect laid the foundation for understanding how CO2 and other gases trap heat within Earth's atmosphere. These researchers, among others, have significantly contributed to our understanding of the intricate connections between CO2 levels and global temperatures. Their studies have been instrumental in establishing the link between human activities and climate change, offering crucial evidence that informs contemporary climate policies and initiatives. Today, their legacy continues to inspire new generations of scientists dedicated to unraveling the complexities of climate dynamics.

The relevance of historical data extends beyond academic interest; it is a critical component of future climate forecasts. By examining past trends, scientists can develop climate models that project future scenarios under various assumptions. These models incorporate historical CO2 levels and temperature records to simulate potential outcomes of continued emissions. Lessons from historical climate reversals, such as the transition from ice ages to warmer periods, guide our understanding of the thresholds and tipping points within the climate system. Recognizing these patterns helps us anticipate potential changes and plan mitigation strategies to prevent catastrophic impacts. The insights gained from historical data underscore the importance of informed decision-making in addressing the challenges posed by climate change. By leveraging this knowledge, we can develop policies and practices that promote sustainability, resilience, and adaptation in the face of a changing climate. Understanding our past is not just an academic exercise; it is a vital step toward securing a sustainable future for generations to come.

Reflecting on these historical perspectives, we see a clear connection between past climate patterns and the challenges we face today. The evidence from the past warns us of the potential consequences of unchecked emissions. Our understanding of historical CO2 fluctuations and their impact on climate serves as a reminder of the delicate balance we must strive to maintain. As we move forward, let us draw on this knowledge to inform our actions and create a more sustainable future.

Just before the publication of this book, Dr. Jim Hansen released the following graph in a Feb. 2025 email called "**The Acid Test: Global Temperature in 2025.**" It is his *Figure 1* for the global temperature compared to a baseline of around 1900. Finding the **Best Linear Fit** starting at 2010 (at +1°C) to be **0.36°C/decade** (as opposed to the 0.18 or 0.24°C/decade NOAA estimates in Chapter 1), we extrapolate that every 30 years another degree C will be experienced globally (e.g., 2040, 2070, 2100, 2130, etc.) but the world emissions seem to take an exponential turn upwards. Only if this global temperature rise is still linear in the future will we see only +5°C by 2130 and climbing (without CDR in the tens of gigatons). Otherwise, it will be even hotter.

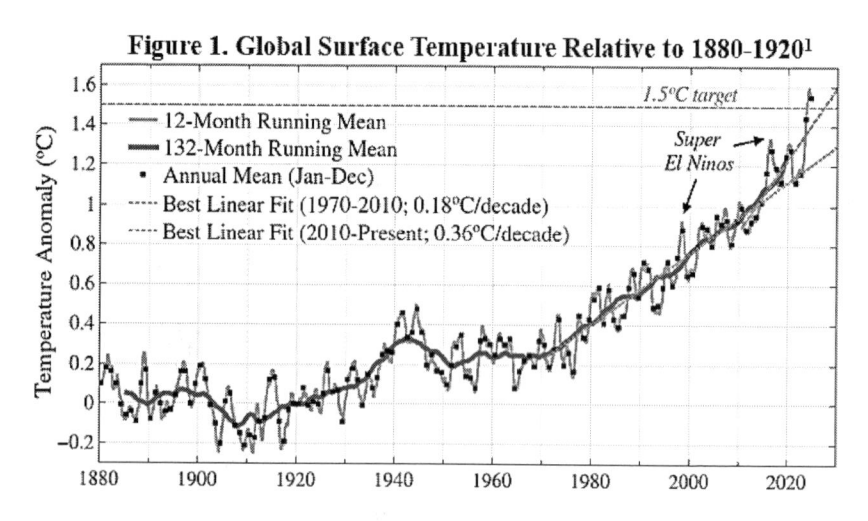

Figure 1. Global Surface Temperature Relative to 1880-1920[1]

This Best Linear Fit for 2010 onward differs from the NOAA graph in Chapter 1 of this book. Dr. Jim Hansen has been precisely predicting our global temperature rise since 1988 with his NASA computer and been on target every time. As we saw in Chapter 1 as well, his accurate prediction in 1988 of the +1.5°C by 2020 was amazing and his projection for 2060 also looks plausible. Join Hansen's Columbia Univ. email list: https://tinyurl.com/HansenEmailList at no cost or obligation.

Reflection Section

Consider how the Vostok Ice Core data informs your understanding of climate change. Reflect on the importance of historical context in shaping modern climate science. How does knowing the past empower us to make informed decisions for the future?

Explore this idea further with a journaling prompt:

- Describe your personal connection to climate change, drawing on historical insights to inform your perspective.
- How does this knowledge shape your view of the present and the choices you make for the future?
- While Hansen relied on the past to create his graph, we found it to be quite useful for estimating the future, so how does the discovery of crocodiles living in the Arctic prepare us for a PETM-style future? (U of Washington, "Climate Change Lecture Notes", www.atmos.Washington.edu)

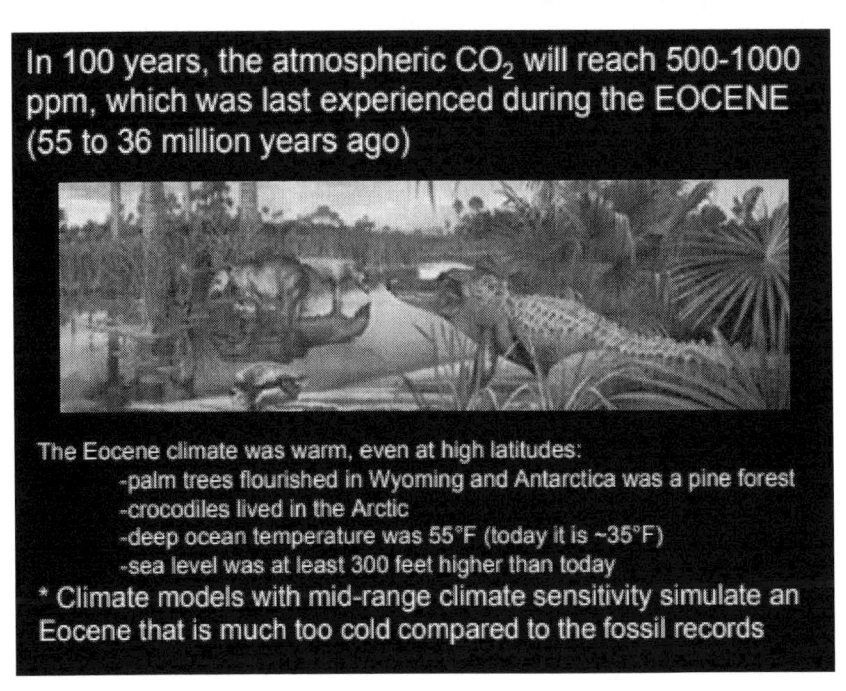

In 100 years, the atmospheric CO_2 will reach 500-1000 ppm, which was last experienced during the EOCENE (55 to 36 million years ago)

The Eocene climate was warm, even at high latitudes:
- palm trees flourished in Wyoming and Antarctica was a pine forest
- crocodiles lived in the Arctic
- deep ocean temperature was 55°F (today it is ~35°F)
- sea level was at least 300 feet higher than today

* Climate models with mid-range climate sensitivity simulate an Eocene that is much too cold compared to the fossil records

Chapter 3

The Gigaton Challenge

Imagine standing on the edge of a vast canyon, where each rock and crevice symbolizes the magnitude of carbon dioxide we must remove to stabilize our climate. This is the challenge we face: a gigaton of CO2, a number so vast that it defies easy comprehension. To put it into perspective, a gigaton equals one billion tons (about equal to a billion metric tonnes). It's a volume that could fill 400,000 Olympic-sized swimming pools or, more dauntingly, it represents the annual emissions of countries like Japan or Germany.

These comparisons illustrate the enormity of our task, yet they also highlight the scale we must achieve to make a significant impact on global emissions, attacking just one gigaton.

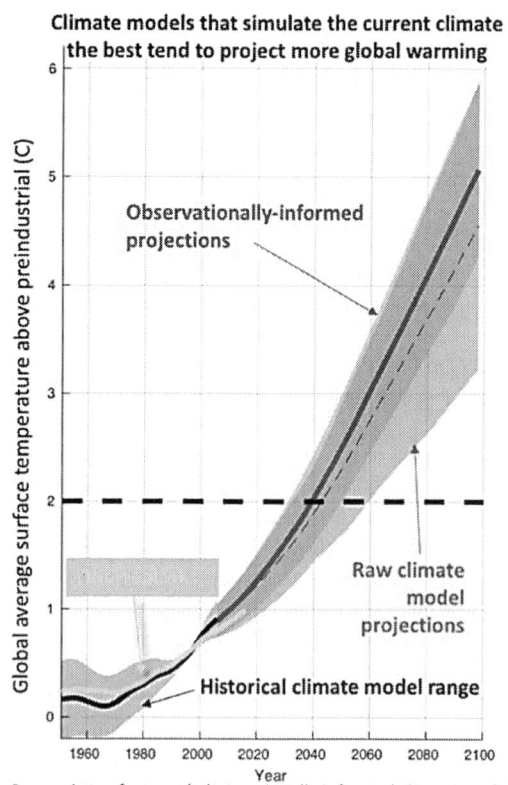

Comparison of raw and observationally-informed climate model projections from Brown and Caldeira (2017, Nature)

The scientific basis for targeting gigaton-level removal is rooted in our understanding of Earth's climate thresholds. Scientists have determined that reducing atmospheric CO2 by several gigatons annually is crucial for keeping global temperatures below 3°C, a target considered vital for avoiding the most catastrophic effects of climate change. Otherwise, the Stanford temperature graph projection from Nature magazine on the previous page will certainly be realized by 2100. By stabilizing CO2 levels, we can halt the relentless march of temperature increases and, in turn, arrest the rise in sea levels. This isn't just about numbers; it's about safeguarding the future of our planet. Addressing this challenge requires a concerted effort, leveraging both technology and nature-based (e.g., solar, wind) solutions to achieve the scale necessary for meaningful change in the amount of heat-blanket CO2.

The potential impact of removing multiple gigatons of CO2 is profound. Climate models project that such reductions could lead to a significant cooling effect, potentially reversing some of the warming trends observed in recent decades. This cooling would also help to stabilize sea levels, protecting coastal communities from the devastating effects of flooding and erosion. Surprisingly, it probably will begin to stabilize the polar ice sheets and encourage their refreezing! It would also contribute to the restoration of ecosystems disrupted by climate change, offering a glimmer of hope for a more sustainable future. Yet, achieving these outcomes requires overcoming substantial technical and logistical barriers, as the infrastructure for carbon capture is not yet sufficient to meet the demand.

The challenges of achieving gigaton-scale removal are formidable. The infrastructure required for carbon capture and storage is immense, demanding vast investments in technology and resources. Energy demands are high, as current methods of CO2 capture are energy-intensive, requiring significant power to operate. Additionally, the allocation of resources must be carefully managed to ensure that carbon removal efforts do not compete with other essential needs, such as food production or energy generation. These obstacles highlight the need for innovation and collaboration as we strive to develop more efficient and cost-effective methods for carbon removal.

3.1 Visualization: Understanding a Gigaton

Visualize a gigaton in practical terms: Imagine a forest the size of a small country, absorbing carbon. Picture the emissions from millions of vehicles disappearing. This scale is what we aim for in carbon removal, a monumental but achievable goal with the right strategies and resources.

Innovation is key to overcoming these hurdles. Advances in technology-based removals, such as direct air capture and bioenergy with carbon capture and storage, offer promising avenues for achieving gigaton-scale impact. These technologies, while currently expensive, are expected to become more affordable as investment and research drive down costs. Nature-based solutions, like reforestation and soil carbon sequestration, also play a critical role, offering lower-cost but equally vital approaches to carbon removal. Together, these strategies form a comprehensive plan for addressing the gigaton challenge, combining the strengths of both technology and nature.

The path forward requires not only technological innovation but also policy support and international cooperation. Governments can facilitate progress by implementing policies that incentivize carbon removal, such as carbon pricing or subsidies for clean technologies. However, the cost (ultimate goal being **$1/ton**) is the biggest hurdle, necessitating the corralling of billionaires and the emerging trillionaires since governments are reluctant to allocate such large sums, even for a single, one-time investment for years of renewable energy powered service. Collaboration between nations, industries, and communities is essential to share knowledge, resources, and best practices. By working together, we can build the infrastructure and systems needed to achieve gigaton-scale carbon removal, paving the way for a more sustainable and resilient future, as renewable substitution for fossil fuels slowly comes online.

3.2 Global Carbon Budgets: Where We Stand

At the heart of climate policy is the concept of a carbon budget, a framework designed to limit global temperature rise by capping the total amount of CO2 that can be emitted without exceeding a specific warming threshold. This concept is crucial because it translates the abstract notion of climate change into tangible, actionable limits. The Intergovernmental Panel on Climate Change (IPCC) plays a pivotal role in defining these budgets and providing guidelines that inform international climate agreements and national policies. The idea is simple: to keep global warming below 3°C, we must not release more than a set amount of carbon into the atmosphere, but this idea is proven to be short-sighted in this book since the driving force for the temperature change is NOT how much CO2 we add today or stop adding to the air but how much ABOVE 290 ppm we have already added since the 1950s. This analysis is based on extensive scientific research that correlates CO2 levels with temperature changes, offering a clear target of around 300 ppm for a livable temperature worldwide, with reduced emissions following up to ensure less CDR will be needed as time goes on.

Analyzing current global emissions against these carbon budgets reveals a sobering reality. As of now, we are emitting CO2 at a rate that far exceeds what these budgets allow if we aim to meet our climate goals. According to the latest data, the world is pumping out approximately 40 billion tons (gigatons) CO2 annually, with major contributors including industrial powerhouses like China, the United States, and India. These emissions are not just numbers; they represent the burning of fossil fuels, deforestation, and industrial processes that collectively drive climate warming leading to a hothouse earth by 2130 (by a linear projection of the present emission rate!). The gap between current emissions and budget targets underscores the urgency of transforming how we produce and consume energy. Without significant reductions, we risk exhausting our carbon budget within a few decades, making it nearly impossible to avoid the most severe impacts of climate change as 5 and 6°C by the end of this century.

This becomes clear from the Hansen graph combined with the Hansen equation. We have passed the 410 ppm mark a couple of years ago. With 290 ppm as our baseline and maximum of CO2 for the past 420,000 years, subtracting the two numbers yields +120 ppm. Using the Hansen equation, we divide 120 by 20 to find how many +degrees of temperature increase we are indebted for, letting history talk sense to us. Most readers of my multiple journal articles react with disbelief to the answer referred to above: **+6°C**. So far, this amount of temperature increase is true today but as we add about 2 ppm (present rate – Hansen, 2025) every year, you can tell it is about one decade to create another degree of temperature increase indebtedness, as the history of the earth has clearly stated in the Hansen graph, unless we also drastically DECREASE our fossil fuel burning! Note the discrepancy between the most recent Hansen projection and the Best Linear Fit of the Global Surface Temperature from Chapter 1 and 2. This is based on the difference between a LINEAR projection and the real EXPONENTIAL reality of the temperature rise worldwide that we are presently experiencing.

One more fact that needs to be shared at this point as the reader attempts to comprehend the import of a few degrees increase:

+6°C is the Extinction Threshold we hope to avoid at all costs!

This should remind the reader to watch the National Geographic video mentioned in Chapter 1 called "Six Degrees Could Change the World" (see https://www.natgeotv.com/ca/six_degrees/about). I personally could not keep viewing it after 3 to 4 degrees since the predictions are increasing dire, visually disturbing and disastrous. **Even at +2°C presently, we are witnessing wildfires everywhere.** Mass migrations from tropical regions will become a major social disruption very soon with another degree or two more. Canada, Greenland, Siberia, and eventually Antarctica will all become the higher priced real estate for human settlements, with equatorial regions *basically uninhabitable*. Keep in mind that during the PETM mentioned earlier, our planet had **alligators in the Arctic**! Furthermore, tropical plant fossils still can be found in Greenland and other Northern areas from the PETM era. Therefore, the pending "extinction" is not an overstatement or an exaggeration when we realize the unavoidable severity of our situation with ever-increasing heat everywhere on the planet due to the excess greenhouse gas (GHG) of CO2 that is ALREADY in the atmosphere (exceeding 290 ppm). Most importantly, the +6°C cited here is only the indebted or stored heat content we presently have created with an excessive 420 ppm of CO2 that the earth has NEVER experienced for over a million years (290 ppm has been the absolute maximum CO2 level). What we need to be aware of even more acutely is emphasized in the text box here.

> Every additional number of indebted degrees C will be added to the +6°C cited here, *one degree for EACH coming decade*, until we increase renewables drastically and at the same time use gigaton CDR for the present global 40 gigatons of annual CO2 emissions.

Today, we read in *New Scientist* that "2/3 of the earth's glaciers will be gone by the end of this century." Despite the daunting challenges, some nations have made commendable progress in aligning their emissions with their carbon budgets. Countries like Sweden and Denmark stand out for their ambitious initiatives, leveraging renewable energy sources to slash their carbon footprints. Sweden, for instance, has invested heavily in wind and solar power, setting a roadmap to become carbon neutral by 2045. Denmark's commitment to offshore wind farms has set a global benchmark for sustainable energy. These success stories demonstrate that it is possible to reconcile economic growth with environmental stewardship. However, not all countries are on track. Many are struggling to meet their targets, hindered by economic dependencies on fossil fuels, political resistance, or lack of infrastructure. This uneven progress highlights the need for increased global cooperation and support, especially for developing nations facing unique challenges which forces them to increase fossil fuel usage at this time.

As our understanding of climate science evolves, so too must our approach to carbon budgeting. New data on emissions trends and climate sensitivity suggest that adjustments may be necessary to keep carbon budgets relevant and effective. For instance, the discovery of previously underestimated feedback mechanisms in the climate system could mean that we have less time than previously thought to achieve our goals. This evolving scientific landscape calls for a dynamic approach to carbon budgets, one that can adapt to new information and changing circumstances. Policy shifts are required to incorporate these insights, ensuring that carbon budgets remain accurate and actionable. Enhanced budget management might involve more stringent emissions targets, increased investment in carbon capture technologies, or more robust frameworks for international collaboration. These adjustments will be critical in guiding global efforts to mitigate climate change and protect our planet for future generations.

3.3 Past Attempts and Lessons Learned

As we explore the landscape of carbon dioxide removal, it's crucial to reflect on the historical efforts that have shaped our current understanding and strategies. Decades of initiatives aimed at reducing carbon emissions have provided a wealth of knowledge, both from their successes and their shortcomings. In the late 20th century, significant attention was directed toward large-scale carbon capture projects. Facilities like Norway's Sleipner Gas Field and Canada's Boundary Dam have been at the forefront, capturing and storing millions of tons of CO_2. These projects have served as important testbeds, revealing both the potential and the limitations of carbon capture technology. While they demonstrated that CO_2 could be effectively stored underground, they also highlighted the economic and technical challenges involved, such as high operational costs and the need for substantial infrastructure investments.

Retrospective evaluations of global policies further enrich our understanding. The Kyoto Protocol, enacted in 1997, marked the first significant international effort to set binding emission reduction targets. While it set an important precedent, the protocol faced substantial hurdles, including a lack of participation from major emitters like the United States and difficulties in enforcing compliance. These challenges underscored the need for more inclusive and flexible agreements that accommodate varying national circumstances. Despite these obstacles, some policies have made notable progress. The European Union's Emissions Trading System, for example, has successfully created a market for carbon allowances, incentivizing reductions across member states. This policy's success demonstrates the effectiveness of market-based solutions in driving emissions reductions, providing valuable lessons for future initiatives.

In examining past attempts at gigaton-scale removal, we can identify key successes and failures. Carbon taxes have emerged as one of the most successful tools, providing a straightforward mechanism to incentivize emissions reductions. Countries like Sweden and British Columbia have implemented carbon taxes that have effectively reduced emissions while maintaining economic growth. These examples show that well-designed fiscal policies can align environmental and financial goals. However, challenges remain, particularly in the deployment of technology. Many carbon capture and storage projects have struggled with scalability and cost-effectiveness, limiting their widespread adoption. This highlights the need for continued innovation and investment to overcome these barriers and make technologies more accessible and economically viable.

Drawing lessons from these efforts is crucial in shaping future strategies. One of the most important insights is the value of international collaboration. Climate change is a global issue that requires coordinated action across borders. Successful initiatives often involve partnerships between governments, industries, and research institutions, leveraging diverse expertise and resources. The Paris Agreement exemplifies this collaborative approach, bringing together nearly every nation to commit to emissions reductions. However, collaboration must go beyond agreements; it requires ongoing dialogue, shared goals, and mutual support to achieve meaningful progress. Another key lesson is the necessity of technological innovation. As the world seeks to meet ambitious climate targets, developing and deploying advanced technologies will be essential. Breakthroughs in direct air capture, which removes CO_2 directly from the atmosphere, and advancements in bioenergy with carbon capture and storage, which combines energy production with carbon sequestration, offer promising pathways to achieve gigaton-scale removal. These technologies have evolved significantly, becoming more efficient and cost-effective, yet further advancements are needed to maximize their potential.

The evolution of carbon removal technologies over time illustrates the dynamic nature of this field. Initial efforts focused on capturing emissions from large industrial sources, such as power plants and refineries. While effective, these methods were limited by their high costs and infrastructure requirements. In response, research and development have led to innovative approaches that expand the scope of carbon capture. Direct Air Capture (DAC) technologies, pioneered by Climeworks.com for example, have made significant strides in utilizing chemical processes to extract CO_2 from ambient air. This method offers the advantage of being location-independent, allowing for flexible deployment across various settings. Similarly, bioenergy with carbon capture has advanced, integrating carbon capture into biomass energy production or carbonate formation for burial to achieve negative emissions. By converting biomass into energy and capturing the resulting CO_2, this approach not only reduces emissions but also provides renewable energy. These technological advancements demonstrate the potential for innovation to overcome past limitations, paving the way for more effective and scalable carbon removal solutions.

3.4 The Urgency of Achieving Gigaton Level of Removal

The urgency of achieving gigaton-level removal of carbon dioxide cannot be overstated. As we move forward, the clock is ticking toward what scientists refer to as climate tipping points. These are critical thresholds in the Earth's system, beyond which the consequences of warming become irreversible and self-perpetuating. Think of it as a domino effect: once a certain point is crossed, the changes may accelerate, leading to severe outcomes that are beyond our control. This is why the timeline associated with carbon removal is not just important—it is critical. Without swift action, we risk entering a phase where the impacts of climate change become increasingly difficult to manage. The longer we delay, the closer we edge toward these tipping points, making the task of stabilization more daunting and costly.

Delaying action on gigaton removal has dire consequences for both the planet's ecosystems and human societies as stated previously. As greenhouse gases continue to accumulate in the atmosphere, the frequency and intensity of extreme weather events are expected to rise, due to more and more heat-trapping of incoming solar radiation. We are already witnessing more frequent hurricanes, wildfires, and floods, each causing widespread devastation and displacing communities. These events are not isolated incidents but part of a broader pattern tied to rising global temperatures. Furthermore, biodiversity faces unprecedented threats as habitats are altered or destroyed. Many species struggle to adapt to the rapid changes, leading to declines in populations and, in some cases, **extinction**. The loss of biodiversity not only disrupts ecosystems but also undermines the services they provide, such as pollination, water purification, and climate regulation. These impacts ripple through human societies, affecting food security, health, and livelihoods, particularly in vulnerable regions.

On the flip side, timely action in implementing gigaton-level removal offers substantial economic and social benefits. Transitioning to green industries can create a wealth of jobs, providing opportunities for employment in sectors like renewable energy, sustainable agriculture, and conservation. These industries not only contribute to reducing carbon emissions but also support economic resilience by fostering innovation and competitiveness. Countries that invest early in these areas are likely to reap the benefits of a green economy, positioning themselves as leaders in sustainable development. Moreover, adopting sustainable practices enhances the resilience of communities, making them better equipped to withstand and recover from climate-related disruptions. By reducing reliance on fossil fuels and improving energy efficiency, societies can achieve greater energy security and decrease their vulnerability to fluctuating energy markets.

IEEE ISTAS 2019   **Sea Encroachment After 2100 in Major Cities**

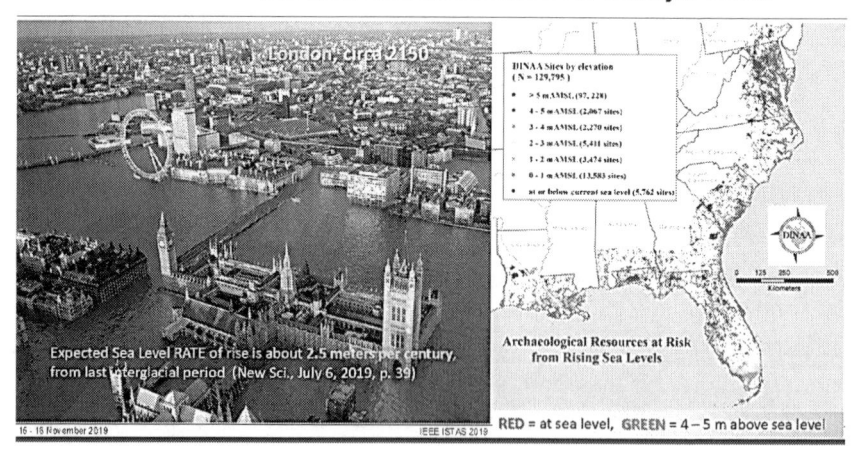

A unified global response is essential in achieving the scale of removal necessary to address climate change. International cooperation enables the sharing of knowledge, technology, and resources, amplifying the impact of individual efforts. Success stories from multinational agreements, such as the Paris Agreement, underscore the potential of collective action. These agreements create frameworks for countries to commit to emissions reductions and support one another in achieving their goals. Enhanced global collaboration can be facilitated through mechanisms like joint research initiatives, technology transfer, and financial support for developing nations. By working together, countries can pool their strengths and overcome the challenges of carbon removal, fostering a sense of shared responsibility and mutual benefit.

The path to achieving gigaton-level carbon removal is complex, requiring coordinated efforts from governments, industries, and communities worldwide or simply a few large CDR plants in various countries. Setting a milestone goal is essential.

Our FIRST GOAL: Achieve 40 gigatons CDR/year

While the challenges are significant, the potential rewards are equally compelling, as we halt the inexorably increasing temperature indebtedness by canceling the annual CO2 emission rate.

IEEE ISTAS 2019 **World Population Growth** – conservative est.

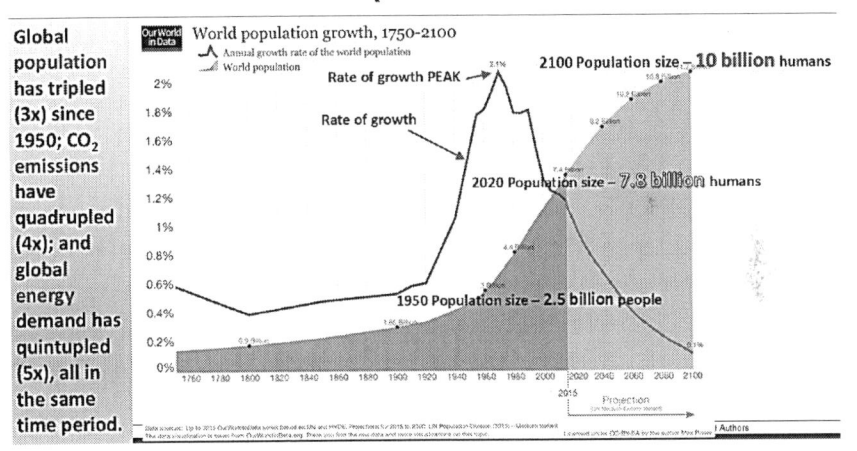

Global population has tripled (3x) since 1950; CO_2 emissions have quadrupled (4x); and global energy demand has quintupled (5x), all in the same time period.

World population growth, 1750-2100
— Annual growth rate of the world population
— World population
Rate of growth PEAK
Rate of growth
2100 Population size – **10 billion** humans
2020 Population size – **7.8 billion** humans
1950 Population size – **2.5 billion** people

The choices we are making today are seen in this illustration on the left side. To understand this comparison with the world population graph of the (1) population size to the (2) rate of growth, I refer you to the Chapter 1 slide of "Carbon Emissions per Annum". There is the change (delta Δ) in CO2 emissions per time. Comparing that with the population "Rate of growth PEAK" here, there is a systems science fact worth noting:

$\Delta CO_2 / \Delta t$

the rate PEAK precedes the size or level peak by *about 100 years*. Therefore, whenever the rate of CO2/time PEAKs, then we can predict with relative certainty that the size or quantity of CO2 in ppm *will level out in 100 years*.

Reflection Section

Many companies are reviewed that are working to make the oceans a better carbon sink in the ***IEEE Spectrum*** (Dec 26, 2024 https://spectrum.ieee.org/ocean-carbon-removal). Some are advocating kelp forests and microalgae in the sea. Another one is Captura's ocean carbon dioxide removal approach, which sucks carbon out of the sea, and ocean alkalinity enhancement, which stores carbon in the sea. "Big funding entities support these ideas. The finalists for the US $100 million XPrize for Carbon Removal …include marine-based strategies, alongside atmospheric ones." However, the magazine emphasizes that "… then there's the issue of scale. To make a dent in the **more than 1,000 gigatonnes of excess CO2 lingering in Earth's atmosphere**, and the few dozen gigatonnes continuing to be emitted each year from human activities, companies would have to process ocean water in biblical proportions." Note how this quote agrees with our estimate in Chapter 4 (where 1100 gigatons = 1000 metric gigatons = 1000 gigatonnes) which really is **ONE TERATONNE** so our task should be called "Teraton CDR Now"!

By prioritizing timely action and international cooperation, we can mitigate the worst impacts of climate change and build a sustainable future for generations to come. As we explore the intricacies of carbon removal and **Carbon Capture & Sequestration (CCS)**, let us remain mindful of the urgency that drives our efforts and the collective power we hold to effect meaningful change. The work begins with the choices we make today. What choices can we make today that will create big strides in gigatonne CO2 burial (CDR)?

Chapter 4

The Role of Billionaires and Trillionaires in Climate Action

In the realm of climate action, billionaires stand as both remarkable figures and contentious subjects. By 2030, the 3 or 4 billionaires mentioned in this chapter *are expected to achieve trillionaire status* (*The Week*, Jan. 2025). Their vast resources and influence allow them to drive significant change, yet their motivations and methods often spark debate. As you navigate this landscape, you might wonder: who are these individuals, and what compels them to invest in environmental causes? Let's delve into the lives and actions of some key players, each contributing uniquely to the fight against climate change. Elon Musk, known for his groundbreaking advancements in renewable energy, is a pivotal figure in this narrative. With a vision that extends beyond our planet, Musk advocates for a future powered by solar energy and hopefully will continue this policy through his Department of Government Efficiency. His work with Tesla's Solar division and the Starlink network demonstrates his commitment to harnessing natural resources for sustainable energy solutions. Musk's belief that a small area in Texas or New Mexico could power the entire United States with solar energy underscores his faith in the potential of renewable sources. But what drives Musk? His motivations are a blend of personal commitment to sustainability and the strategic foresight to tap into future markets that prioritize clean energy.

Jeff Bezos, another titan of industry, has made waves with his ambitious Earth Fund. Launched in 2020, this fund aims to distribute $10 billion to tackle climate change and biodiversity loss. With over 230 grants dispensed, Bezos has funneled significant resources into initiatives ranging from AI environmental solutions to clean energy projects in underprivileged areas. Despite criticism and concerns over potential conflicts of interest, Bezos' contributions remain a significant force in the climate sector. His motivations seem to intertwine personal dedication to environmental stewardship with a business acumen that recognizes the shifting tides of market demand. The Earth Fund's involvement in carbon offset initiatives reflects a strategic approach to influence and shape emerging sectors within the climate economy.

Bill Gates, a name synonymous with innovation, has turned his focus to carbon capture technologies. Through his Breakthrough Energy initiative, Gates has invested $40 million in testing direct air capture technologies in Canada. This investment highlights his commitment to accelerating the deployment of technologies capable of removing billions of metric tons of CO_2 from the atmosphere. Gates' motivations are rooted in a desire to reduce the "green premium," which represents the additional cost of choosing sustainable options over traditional ones. By supporting early-stage climate tech firms, Gates leverages his influence to drive innovation and make sustainable solutions more accessible. This approach not only aligns with his personal values but also positions him as a leader in shaping the future of climate technology.

Richard Branson, the charismatic founder of the Virgin Group, has made his mark with the Virgin Earth Challenge. Though the competition itself was discontinued without a winner, it set a precedent for incentivizing innovation in greenhouse gas removal. Branson's $25 million prize aimed to inspire commercially viable solutions capable of removing significant amounts of carbon from the atmosphere annually. His motivations are multifaceted, encompassing a genuine concern for the planet's future and an understanding of the potential benefits these innovations could bring to global markets. Branson's efforts highlight the role of entrepreneurship in addressing climate challenges, showcasing how business interests can align with environmental goals.

The scope of investments by these billionaires is vast, spanning renewable energy projects, carbon offset programs, and cutting-edge technologies. Their financial contributions often surpass those of governmental bodies, influencing market trends and driving the adoption of sustainable practices. In comparison to state-funded initiatives, billionaire investments can be more agile, allowing for rapid response to emerging opportunities in climate innovation. This agility, combined with their willingness to take risks on unproven technologies, positions these individuals as catalysts for change. Yet, their influence also raises questions about accountability and the potential for these investments to shape the priorities of the climate sector.

Researchers at ETH Zurich have found that "To slow the pace of global warming, we need to drastically reduce greenhouse gas emissions. Among other things, we need to do without fossil fuels and use more energy efficient technologies, the researchers say. However, reducing emissions alone won't do enough to meet the climate targets. Large quantities of carbon dioxide must be captured from the atmosphere and either stored permanently underground or used as a carbon neutral feed stock in industry. Unfortunately, the carbon capture technologies available today require a lot of energy and are correspondingly expensive. which is why the researchers at ETH Zurich are developing a new method that uses light. In the future, the energy required for carbon capture using this process will come from the sun." CleanTechnica.com Dec. 2023; https://doi.org/10.1021/acs.chemmater.3c02435

Recently, IEA published a guidebook for Direct Air Capture (DAC) https://iea.blob.core.windows.net/assets/78633715-15c0-44e1-81d

4.1 Visualization: Billionaires' Impact on Climate Action

To begin with, some may ask, "Why recruit billionaires and trillionaires instead of government funding sources?" The answer is simple. Not a single government around the world has been successful in reducing their carbon emissions significantly to achieve "negative emissions," that is where more carbon is absorbed from the air and environment than is emitted. Google tells us "According to current information, no government has officially achieved "negative emissions" on a large scale; however, some countries like Bhutan and

Suriname are considered to be "carbon negative" due to their large forested areas that absorb more carbon dioxide than they emit, effectively achieving a net negative emission status." Furthermore, the answer is cost. We find that due to this author's further analysis in creating *an annotated enhancement to the Hansen Graph* online at https://tinyurl.com/SixDegreeGraph, even using the ten-year-old CO_2 level of 410 ppm minus the pre-industrial 290 ppm yields about 120 ppm of excess CO_2 in the atmosphere. Converting this amount into gigatons of CO_2 we simply **multiply by 7.77 Gt/ppm** to find the earth is indebted for about **932 Gt** of CO_2 that MUST be removed by carbon capture and CDR on a gigaton level, probably over decades, in order to reverse the heat-trapping overreach of CO_2, as revealed *quantitatively by the Hansen Graph*. In just a few more years, this will amount to **over a trillion tons of CO_2** (1000 Gt) CDR when it finally comes online in earnest and at least $1 trillion at $1/ton.

> However, we can project with high confidence a reassuring prediction for only *150 Gt of Direct Air Capture* and storage of CO_2 underground in a carbonate rock, we will actually realize a 1^0C DECREASE worldwide in our indebted temperature burden.

Imagine a chart illustrating the financial contributions of billionaires versus governmental funding in climate initiatives. This visual representation can help underscore the magnitude of their investments and their potential to drive significant change in the sector. It will also benefit their investments to drive down CO_2 levels. A comprehensive McKinsey Report details how gigaton CDR can be done: **"Carbon removals: How to scale a new gigaton industry"** (https://www.mckinsey.com/capabilities/sustainability/our-insights/carbon-removals-how-to-scale-a-new-gigaton-industry)

As you consider the role of billionaires in climate action, it's important to recognize both the potential benefits and the complexities of their involvement. Their contributions can drive innovation and accelerate progress, but they also invite scrutiny and debate. Understanding their motivations and the scope of their investments provides a nuanced perspective on how these influential figures shape the future of our planet.

4.2 Billionaire-Led Innovations in Carbon Capture

In the quest to tackle climate change, billionaire-backed innovations have been pivotal, particularly in the realm of carbon capture technologies. These cutting-edge solutions are pushing the boundaries of what's possible in removing carbon dioxide from our atmosphere. One area of significant development is enhanced direct air capture (DAC) systems. These systems aim to extract CO2 directly from the air, offering a promising way to address emissions that are otherwise difficult to mitigate. Enhanced DAC technologies are gaining traction thanks to substantial investments, allowing for improvements in efficiency and scalability. With funding in place, companies can experiment with novel materials and processes to capture CO2 more effectively. These innovations are not just theoretical; they are being tested in real-world environments, setting the stage for broader adoption.

Ocean fertilization experiments represent another frontier in billionaire-funded climate action. This technique involves adding nutrients to ocean waters to stimulate the growth of phytoplankton, which absorb CO2 during photosynthesis. The concept is straightforward, yet the execution is complex, requiring careful consideration of ecological impacts. Despite these challenges, the potential to draw down large quantities of carbon makes ocean fertilization an attractive option. With financial backing, scientists and researchers are able to conduct experiments that explore the feasibility and sustainability of this approach. By understanding the ecological dynamics involved, these projects aim to ensure that carbon sequestration does not come at the expense of marine health.

Advanced bioenergy systems, particularly those integrating carbon capture and storage (CCS), are also receiving attention and investment. These systems convert biomass into energy while capturing the resulting CO2 emissions, offering a dual benefit of renewable energy production and carbon reduction. The integration of CCS with bioenergy is seen as a pathway to achieving negative emissions, where more carbon is removed from the atmosphere than is emitted. By investing in these technologies, billionaires are supporting the development of systems capable of operating on a large scale, paving the way for significant contributions to global carbon reduction targets.

Collaboration with research institutions is a key element in the success of these innovations. Billionaires often partner with universities and independent research labs to drive technological advancements. These partnerships provide the technical expertise and research capabilities needed to bring ambitious projects to fruition. Joint ventures with academic institutions allow for access to cutting-edge research and a pool of talented scientists and engineers. By funding independent labs, billionaires can support exploratory projects that have the potential to yield breakthrough technologies. This collaborative approach not only accelerates innovation but also ensures that projects are grounded in rigorous scientific research.

The scalability of these innovations is crucial for achieving meaningful impact. Pilot projects serve as testing grounds where technologies can be refined and optimized before being deployed on a larger scale. Transitioning from pilot to full-scale operations involves overcoming technical, financial, and regulatory hurdles. High initial costs are a significant barrier, as the infrastructure required for large-scale deployment can be expensive. However, as technologies mature and production processes become more efficient, costs are expected to decrease. Regulatory approval processes also pose challenges, as new technologies must meet stringent environmental and safety standards. Navigating these processes requires collaboration with regulatory bodies to ensure compliance and build public trust.

Deploying these innovations at a gigaton scale presents additional challenges. Technical obstacles include improving the efficiency and reliability of carbon capture systems to handle larger volumes of CO_2.

Financial hurdles involve securing the necessary investments to scale up operations while maintaining economic viability. Overcoming these challenges requires a concerted effort from all stakeholders, including governments, businesses, and research institutions. By addressing these obstacles, billionaire-backed projects can pave the way for the widespread adoption of carbon capture technologies, contributing significantly to global efforts in combating climate change.

4.3 Success Stories: Billionaire-Funded Climate Projects

In the realm of climate action, certain projects stand out not only for their ambition but for their tangible successes in reducing carbon emissions. Among these is Tesla's Gigafactory, a monumental facility dedicated to producing batteries for electric vehicles and energy storage systems. Located in Nevada, the Gigafactory represents a significant leap forward in renewable energy storage, a critical component in the transition away from fossil fuels. By producing batteries at scale, Tesla aims to reduce the cost of energy storage, making renewable energy more accessible and viable on a large scale. This project has been instrumental in accelerating the adoption of electric vehicles, which are crucial for reducing emissions from the transportation sector. The Gigafactory's success lies not just in its technological advancements but in its strategic partnerships with local governments. These collaborations have provided essential support in terms of incentives and infrastructure, paving the way for successful implementation and expansion.

Another example is the Bezos Earth Fund's reforestation initiatives. These projects aim to restore degraded landscapes, enhancing biodiversity while capturing carbon dioxide from the atmosphere. By focusing on reforestation, the Earth Fund addresses two critical issues: reversing deforestation and mitigating climate change. The success of these initiatives can be attributed to strong community engagement and support. Local populations are often involved in the planning and execution of reforestation efforts, ensuring that the projects align with regional needs and priorities. This grassroots involvement not only enhances the effectiveness of the reforestation process but also fosters a sense of ownership and responsibility among local communities. Additionally, the collaboration with environmental organizations and scientific bodies ensures that the reforestation practices are sustainable and ecologically sound.

The broader implications of these projects extend beyond their immediate impact. Tesla's Gigafactory has set a new benchmark for the energy storage industry, inspiring similar initiatives globally. By proving that large-scale battery production is both feasible and profitable, Tesla has encouraged other companies to invest in renewable energy solutions. This ripple effect can lead to increased innovation and competition, ultimately driving down costs and expanding access to clean energy technologies. Similarly, the Bezos Earth Fund's reforestation efforts have highlighted the importance of nature-based solutions in climate strategies. By demonstrating the potential of reforestation to capture carbon and restore ecosystems, these initiatives provide a model for other organizations and governments seeking to enhance their climate efforts.

The lessons learned from these successful projects offer valuable insights for future billionaire-funded climate initiatives. A key takeaway is the importance of transparency and accountability. By openly communicating their goals, processes, and outcomes, these projects build trust with stakeholders and the public. This transparency ensures that funds are used effectively and that progress can be accurately assessed and reported. Moreover, adaptive management strategies have proven essential in navigating the complexities of climate action. By remaining flexible and responsive to changing conditions, these projects can adjust their approaches as needed, ensuring sustained impact and relevance. This adaptability is crucial in a field where scientific understanding and technological capabilities are continually evolving.

As we look to the future of climate action, the success stories of billionaire-funded projects remind us of the potential for innovation and collaboration to drive meaningful change. They illustrate how strategic partnerships, community engagement, and a commitment to transparency can enhance the effectiveness of climate initiatives, setting the stage for a more sustainable and resilient world.

4.4 Accountability and Transparency in Climate Funding

In the landscape of climate philanthropy, accountability and transparency are not just buzzwords; they are the bedrock of trust and effectiveness. When billionaires pledge vast sums to environmental causes, the world watches closely. Ensuring that these funds reach their intended projects is crucial for maintaining the integrity of these initiatives. Without clear oversight, even the most generous contributions can fall prey to inefficiencies or mismanagement. By making their funding processes transparent, philanthropists can build public trust, demonstrating that their intentions are not only noble but also actionable and impactful. Public trust is essential, as it encourages more stakeholders to engage with and support these initiatives, creating a ripple effect of positive change.

Currently, practices around transparency in climate funding vary significantly among billionaires. Some philanthropists have taken steps to openly report how and where their funds are allocated. Public reporting mechanisms, such as annual reports and financial disclosures, serve as vital tools in this endeavor. They allow both the public and stakeholders to track the progress and efficacy of funded projects. However, not all contributions are equally transparent. Inconsistencies in reporting can lead to skepticism, as stakeholders may question whether these funds are being used effectively. When billionaires fail to provide clear and detailed information about their investments, they risk undermining the very projects they aim to support.

Third-party assessments and audits play an indispensable role in maintaining accountability. Independent evaluations are crucial in providing an unbiased perspective on the effectiveness and impact of climate funding. External audits scrutinize financial flows and project outcomes, ensuring that funds are utilized as intended. Certification by environmental watchdogs offers an additional layer of credibility, validating the claims made by philanthropists regarding their environmental contributions. By engaging third parties in the evaluation process, philanthropists can demonstrate a commitment to transparency and accountability, reinforcing public confidence in their efforts. This independent oversight not only enhances the credibility of these initiatives but also ensures that they deliver tangible results.

Yet, there is ample room for improvement in transparency practices. One proposed strategy is the adoption of standardized reporting frameworks that can provide a consistent and uniform method for disclosing climate investments. Such frameworks would enable philanthropists to report their contributions clearly and comparably, simplifying the process for both donors and recipients. Additionally, engaging with community stakeholders is vital in enhancing transparency. By involving local communities in the decision-making process, philanthropists can ensure that projects align with regional needs and priorities. This collaboration fosters mutual trust and accountability, as it allows stakeholders to provide feedback and hold philanthropists accountable for their commitments.

As the chapter on the role of billionaires in climate action draws to a close, the importance of accountability and transparency becomes clear. These principles are not mere formalities but essential components of effective climate philanthropy. By embracing transparency, billionaires can maximize the impact of their contributions, ensuring that they drive meaningful change in the fight against climate change. As we transition into the next chapter, we will explore how innovative technologies and community-driven initiatives can further amplify the efforts of these philanthropic giants, paving the way for a more sustainable and resilient future.

Chapter 5

Innovative Technologies for Carbon Capture

Imagine a world where the air we breathe can be cleansed of its excess carbon dioxide, much like a filter purifies water. This is the promise of **Direct Air Capture (DAC)**, a groundbreaking technology that addresses the challenge of rising CO2 levels head-on. Unlike traditional methods that capture emissions at the source, DAC systems extract CO2 directly from the ambient air, offering a versatile solution to our climate crisis. Think of it as a giant vacuum cleaner for the atmosphere, working tirelessly to pull carbon from the sky and either store it safely underground or repurpose it for industrial use. The process begins with large fans drawing in air, which then passes over materials that bind to carbon dioxide molecules through chemical reactions. These materials, often composed of amines or hydroxides, are adept at selectively absorbing CO2 while allowing other gases to pass through. Once saturated, the materials release the captured carbon when heated, allowing it to be collected and stored or utilized.

The scalability of DAC and CDR technology is a topic of intense discussion and research. As of now, DAC plants worldwide capture about 11,000 tons of CO2 annually, but the potential for growth is immense. The International Energy Agency projects that by 2050, **DAC could reach a capacity of over a gigaton per year**. However, scaling up to this level requires addressing several technical and economic challenges. Pilot projects, such as those by Climeworks and Carbon Engineering, are paving the way. **Climeworks' Orca plant** in Iceland, for instance, captures 4,000 metric tons of CO2 per year and serves as a crucial testbed for refining DAC processes. These projects demonstrate that while current outputs are modest, the foundation for significant expansion exists. The energy requirements for DAC are substantial, as the process involves both capturing and releasing

CO2, which necessitates heat and electricity. Efficiency metrics are improving, with companies striving to reduce the energy input needed per ton of captured carbon, thus making the technology more viable at larger scales, especially if renewable energy is utilized.

Several companies are in charge of advancing DAC technology, each contributing unique innovations and approaches. **Climeworks**, with its Orca plant, is a pioneer in the field, utilizing a modular system that can be scaled to meet increasing carbon capture demands. **Carbon Engineering**, another leader, is developing a facility in Texas with a projected capacity <u>of capturing one million tons of CO2 per year</u>. Their approach focuses on integrating DAC with existing industrial processes, creating a seamless system that enhances efficiency. These companies, along with others like CarbonCapture Inc. and Global Thermostat, are setting the stage for DAC to become a mainstream solution in the fight against climate change. Each project provides valuable insights into optimizing the technology, from improving the materials used to enhancing the design of capture units.

Despite its promise, DAC faces significant economic and environmental challenges. The initial capital investment required to set up DAC facilities is considerable, often running into millions of dollars. Operational costs add another layer of complexity, as maintaining and running these plants demands a continuous energy supply. Overcoming these financial barriers is crucial for widespread adoption. Additionally, large-scale deployment of DAC has implications for land use and resource allocation. The infrastructure needed for DAC plants must be carefully planned to minimize environmental impact and avoid competing with other essential land uses. Balancing these factors requires thoughtful consideration and strategic planning, ensuring that DAC contributes positively to climate goals without unintended consequences.

5.1 Visualization: DAC in Action

Imagine a sprawling landscape dotted with sleek, futuristic structures—these are DAC plants, tirelessly working to clean our atmosphere. Picture a chart showcasing the potential growth of DAC capacity from current levels to projected gigaton-scale operations by 2050; this visual underscores the transformative potential of the technology.

Direct Air Capture holds immense promise for addressing the climate crisis. Its ability to capture CO_2 directly from the air, independent of emission sources, makes it a versatile tool in our arsenal against global warming. While challenges remain, the innovations and dedication of leading companies offer a path forward.

5.2 Ocean-Based Carbon Sequestration Methods

Oceans, vast and teeming with life, have long played a crucial role in regulating Earth's climate. As natural carbon sinks, they absorb and store substantial amounts of carbon dioxide from the atmosphere. Ocean-based carbon sequestration methods aim to enhance this natural process, providing another avenue to mitigate climate change. One such method is ocean fertilization, which involves adding nutrients like iron powder to ocean waters. This stimulates the growth of phytoplankton, microscopic plants that absorb CO_2 during photosynthesis. As these organisms flourish, they draw more carbon from the atmosphere, and when they die, the carbon is sequestered in the deep ocean. Another promising technique is alkalinity enhancement, which involves introducing alkaline substances—such as crushed limestone—into seawater. This process increases the ocean's capacity to absorb CO_2 by neutralizing the acidity of the water, allowing it to store more carbon in the form of **dissolved bicarbonates.**

While these methods hold potential, they also come with risks and challenges. **Ocean Fertilization** with iron filings, is promoted by the Woods Hole Oceanographic Institution which states, "Iron fertilization is a Carbon Dioxide Removal (CDR) technique that would artificially add iron to the ocean's surface to stimulate growth of phytoplankton." Some estimates suggest that fertilization using macronutrients could be cheaper (around $20 per ton of CO2) but are less scalable, and monitoring costs are excluded. The sudden influx of nutrients will lead to algal blooms however, which can deplete oxygen levels but also trap CO2 in the *range of millions of tons*. However, some say this disruption could affect biodiversity, alter food webs and impact species that rely on stable environmental conditions, even though algae blooms are common in the ocean historically from the Sahara Desert winds.

Our Integrity Research Institute sponsored Russ George from Planktos Corp. in 2006 at COFE3 to present his pioneering trips to throw iron filings into the ocean and monitor the plankton blooms afterwards. He showed satellite images of the green plankton blooms and the tests of them falling below the thermocline to the ocean bottom, thus capturing CO2 with them. Since then, many experiments have been performed. One report in *New Scientist* (July 18, 2012) heralded a success with the title, "Geoengineering with iron might work after all." The article describes the tiny floating algae (phytoplankton) pulling carbon dioxide out of the atmosphere. "When they die, the plankton sink to the seabed, taking the carbon with them. Over thousands of years, this strips CO2 from the air, lowering temperatures... But many ocean regions are short of iron, which plankton need to grow, so the process does not occur. Adding iron should stimulate plankton growth in these areas." The Alfred Wegener Institute for Polar and Marine Research in Bremerhaven, Germany, and colleagues dumped iron sulphate into a slowly rotating eddy's core 60 kilometers across and studied the resulting bloom, dominated by diatoms, for a trial called Eifex. "The diatom bloom grew for three weeks, then died and sank. At least half of it sank far below 1 kilometre, and probably reached the sea floor" (*Nature,* DOI: 10.1038/nature11229). Further encouragement came from the same scientist who coauthored the composite climate model graph projecting temperature to 2100, as shown in Chapter 3. The article concludes with the statement, "At most, a global programme could mop up <u>about 1 gigatonne of carbon per year</u>... according to a modelling study by **Ken Caldeira** at the Carnegie Institution of Washington in Stanford, California (*Climatic Change*, DOI: 10.1007/s10584-010-9799-4). "It's too little to be the solution...but it's too much to ignore."

On the other hand, alkalinity enhancement appears to offer a more stable solution, but its long-term effects on marine life remain uncertain. Both techniques aim to increase the ocean's carbon storage potential, yet they require careful consideration of ecological impacts. Ensuring that these methods do not inadvertently cause more harm than good is paramount. The promise of long-term carbon storage through these methods is enticing, but it must be balanced with a commitment to preserving ocean health. Research and development in ocean-based sequestration are ongoing, with several significant projects underway. **Ocean Alkalinity Enhancement projects,** mentioned in Chapter 3, are exploring how naturally occurring alkaline minerals can increase the ocean's carbon-capturing ability. Researchers are conducting experiments to determine the optimal types and quantities of minerals to use, aiming to maximize effectiveness while minimizing environmental disruption. Iron fertilization experiments are also progressing, with scientists investigating the best ways to introduce iron without triggering harmful algal blooms. These initiatives are crucial for advancing our understanding of how ocean-based methods can be safely and effectively implemented. By gathering data from these projects, scientists hope to refine the techniques and develop guidelines for their responsible use.

Regulatory and ethical considerations are integral to the deployment of ocean-based carbon sequestration methods. International maritime laws govern activities that could impact marine environments, ensuring that any interventions are subject to rigorous scrutiny. The London Protocol, for instance, regulates ocean fertilization, requiring comprehensive assessments before any large-scale projects can proceed. Ethical debates surrounding geoengineering and the deliberate modification of Earth's systems are also prevalent. Proponents argue that these methods are necessary to combat climate change, while critics caution against unintended consequences that could arise from manipulating natural processes. Balancing these perspectives is vital for developing frameworks that safeguard both the environment and ethical standards. As we explore these technologies, it is crucial to weigh the potential benefits against the moral and ecological implications, ensuring that we proceed with caution and responsibility.

5.3 Advanced Bioenergy with Carbon Capture and Storage (BECCS)

In the realm of carbon capture, Advanced Bioenergy with Carbon Capture and Storage, or BECCS, stands as a promising ally. This technology merges two critical processes: bioenergy production and carbon capture. Let's break it down. Bioenergy production involves using organic materials, known as biomass, as a feedstock. Think of this as turning plant matter or agricultural waste into fuel. The beauty of biomass is its ability to regrow, making it a renewable resource. When we burn this biomass for energy, CO_2 is released, but here's where the carbon capture comes in. Carbon capture technologies trap the released CO_2 before it can escape into the atmosphere. This captured CO_2 is then stored underground, effectively removing it from the carbon cycle. The result? A process that not only generates energy but also has the potential to be carbon-negative, meaning it removes more CO_2 than it emits.

The potential for carbon-negative energy production through BECCS is not just theoretical. Real-world examples provide proof of concept. Take, for instance, the **Drax Power Station** in the UK, which has integrated BECCS technology into its operations. By capturing and storing significant amounts of CO_2, Drax aims to become the world's first carbon-negative power station. This achievement is significant because it demonstrates how BECCS can outperform traditional energy sources, which typically add CO2 to the atmosphere. Compared to fossil fuel-based power generation, BECCS offers a dual benefit: it provides energy while actively reducing atmospheric CO2 levels. This dual capability positions BECCS as a formidable option in our arsenal against climate change, providing both a renewable energy source and a means to mitigate emissions.

Yet, like all innovative technologies, BECCS faces hurdles. One major challenge is land use competition. Biomass cultivation requires land, which could otherwise be used for food production. This creates a delicate balance between energy needs and food security. In regions where arable land is limited, the choice between growing crops for energy versus food becomes a critical issue. Additionally, the energy input versus output efficiency of BECCS is an ongoing topic of research. The process of capturing and storing CO2 demands energy, and ensuring that the energy produced exceeds the energy consumed is essential for the viability of BECCS. These technical and ecological challenges are compounded by social considerations, such as public acceptance and regulatory frameworks, which vary across regions.

Innovation is the key to overcoming these challenges, and recent advancements in BECCS technology show promise. Researchers are developing high-yield biomass species that can grow quickly and efficiently, maximizing the energy produced per acre of land. These species are bred to thrive in diverse climates, reducing the need for extensive agricultural inputs. On the carbon capture front,

---

Case Study: Drax Power Station

Imagine a power station that doesn't just stop at generating energy but takes it a step further by actively removing carbon from the atmosphere. The Drax Power Station in the UK is doing just that. By integrating BECCS technology, Drax captures and stores large volumes of CO2, setting a precedent for carbon-negative energy production. This case study illustrates the real-world application of BECCS and its potential to transform the energy landscape.

improvements in capture methods are making BECCS more efficient and cost-effective. New capture technologies are being designed to operate at lower temperatures and pressures, reducing the energy required for CO_2 capture and storage. These advancements not only enhance the overall efficiency of BECCS but also make it more scalable and adaptable to different regions and energy needs.

As we explore the potential of BECCS, it's clear that its integration of bioenergy and carbon capture offers a viable path to reducing net CO_2 emissions. While challenges remain, the ongoing innovations and breakthroughs in BECCS technology hold promise for a future where energy production contributes positively to our climate goals.

5.4 The Future of Mineral Carbonation

Mineral carbonation offers an intriguing pathway for carbon dioxide storage by leveraging the natural chemical reactions between minerals and CO_2 to form stable carbonates. Picture vast tracts of basaltic rock, rich in minerals like olivine and serpentine, reacting with CO_2 to create solid, stable compounds. This process mirrors natural weathering but accelerates it, converting gaseous carbon dioxide into solid carbonate minerals over time. The chemical reactions typically involve CO_2 reacting with silicate minerals to produce magnesium or calcium carbonates, both of which are stable for thousands of years. These reactions occur naturally, albeit slowly, as rainwater interacts with rocks, but when engineered, they can be expedited to sequester carbon on a meaningful scale.

The potential scale of mineral carbonation is vast, with estimates suggesting that suitable rocks could store trillions of tons of CO_2 globally. The permanence of these carbonate formations is a significant advantage, offering a long-term solution to carbon storage without the risk of leakage. Once CO_2 is converted into solid carbonates, it remains locked away indefinitely, effectively removing it from the atmospheric cycle. This stability is crucial, as it ensures that the carbon dioxide sequestered today does not become a problem in the future. However, scaling this process to meet global needs involves substantial logistical and economic challenges. The feasibility lies in mining and transporting the required minerals to sites where carbonation can occur efficiently, a task that demands careful planning and resource management.

Current research and pilot projects are exploring the viability of mineral carbonation, with promising experiments underway. Projects utilizing olivine and basalt rocks are at the forefront, leveraging these abundant minerals for carbon sequestration. The **CarbFix** project in Iceland is a prime example, where CO2 is dissolved in water and injected into basalt formations, converting it into solid carbonates within months. University-led initiatives are also delving into the potential of different mineral types and reaction conditions to optimize the process. These studies aim to refine the techniques used, ensuring that mineral carbonation can be applied widely and effectively. By experimenting with various minerals and conditions, researchers are unlocking new insights into how to accelerate and scale this natural process.

ProjectVesta.org → Gigaton CCS

- Carbon Capture and Storage (CCS)
- Chevron.com/possibilities: 4 megatons/year
- Vesta CEO projects cost at $10/ton

How green sand could capture billions of tons of carbon dioxide
Scientists are taking a harder look at using carbon-capturing rocks to counteract climate change, but lots of uncertainties remain.

by James Temple   June 22, 2020

MIT Technology Review

Rolling stones; fast weathering of olivine in shallow seas for cost-effective CO₂ capture and mitigation of global warming and ocean acidification
R. D. Schuiling and P. L. de Boer

Harnessing Nature
Project Vesta's approach dramatically accelerates Earth's natural longterm CO2 removal process. We make green-sand beaches with a highly abundant volcanic mineral, olivine. We acquire nearby olivine and transport it to beaches where wave action speeds up the carbon dioxide capture process, while also de-acidifying the ocean.

One of the companies dedicated to gigaton carbon capture is **Project Vesta**, which has as its website motto, "Harnessing the power of the oceans to remove excess CO_2 from the atmosphere." Visit https://www.vesta.earth/ for more information about their exciting technique under evaluation in the environment at this moment. Carbon-removing sand made of the ground-up mineral **olivine** is added to the ocean. There, the sand dissolves, countering ocean acidification and permanently removing carbon dioxide from the atmosphere. Another great company with a different approach is **CarbonEngineering.com** in Canada. They use Direct Air Capture which is a technology that captures carbon dioxide directly from the air with an engineered, mechanical system. Direct Air Capture (DAC) technology does this by pulling in atmospheric air, then through a series of chemical reactions, extracts the carbon dioxide (CO2) from it while returning the rest of the air to the environment. This is what plants and trees do every day as they photosynthesize, except Direct Air Capture technology does it much faster, with a smaller land footprint, and delivers the carbon dioxide in a pure, compressed form that can then be stored underground, whereas trees will eventually return the CO2 to the air when they die and decay.

Despite its promise, mineral carbonation faces economic and logistical hurdles. Mining the necessary minerals and transporting them to carbonation sites can be costly, as can the energy required to accelerate the reactions. These processes demand significant upfront investments, which can be a barrier to widespread adoption. Energy input is another concern, as the reactions often require elevated temperatures and pressures to proceed at a practical pace. Balancing the energy costs with the benefits of carbon sequestration is a key challenge that researchers and developers are working to address. Innovative approaches are needed to reduce these costs, making mineral carbonation a viable option for large-scale deployment.

Looking to the future, several innovations hold promise for enhancing mineral carbonation. One exciting area of development is the use of industrial byproducts, such as steel slag or fly ash, as alternative sources of reactive minerals. These byproducts, often considered waste, can be repurposed for carbonation, reducing the need for fresh mining and cutting costs. Advancements in reaction acceleration techniques also offer hope, with researchers exploring methods to speed up carbonation without excessive energy inputs. By refining catalysts and optimizing reaction conditions, it's possible to make mineral carbonation more efficient and economically feasible. These innovations have the potential to transform mineral carbonation into a cornerstone of global carbon management strategies, contributing significantly to our climate goals.

Reflection Section

As we close this chapter on innovative carbon capture technologies, the case has been made that each method, from direct air capture to mineral carbonation, plays its unique role in addressing the pressing need to reduce atmospheric CO2. These technologies, though varied in approach, all aim to mitigate the effects of climate change and restore balance to our planet's climate systems. See what you believe will be the deciding factor in creating a scalable gigaton CDR, critical in scaling these solutions and ensuring their successful implementation on a global scale. Also, <u>contemplate and visualize</u> that by the time several degrees have been equalized in the very near future by the appropriate number of gigatons of CDR, nicely reduced by Carbon Capture and Storage (CCS)...

...a cooling trend will start to become a welcome worldwide phenomenon, which realistically and surprisingly will occur in only a few decades, *in the same amount of time it took to increase the global temperature by the same number of degrees now in reversal.*

Chapter 6

Community-Driven Initiatives and Success Stories

Imagine standing at the forefront of a movement, your voice echoing alongside countless others united in a chorus for change. This is the essence of grassroots environmental movements, where individuals and communities coalesce to drive substantial environmental transformation from the ground up. In today's world, where climate change poses unprecedented challenges, these movements have emerged as powerful catalysts for action. They are not just about protest; they are about creating a dialogue, influencing policy, and reshaping public awareness.

6.1 Grassroots Movements: Power from the Bottom Up

One of the most compelling examples of grassroots activism is the Sunrise Movement. Originating in the United States, this youth-led movement has become a formidable force in climate advocacy. By championing the **Green New Deal**, a comprehensive policy proposal aimed at tackling climate change and economic inequality, the Sunrise Movement has succeeded in shifting the national climate agenda. Their methods are as innovative as they are impactful. In 2018, the movement gained significant attention with a sit-in at Speaker Nancy Pelosi's office, an event that underscored their commitment to non-violent protest and strategic media engagement. This approach has expanded their influence, leading to the establishment of 290 hubs across the country. Their work emphasizes a crucial intersection between climate action and social justice, aligning environmental goals with broader issues of racial and economic equality.

Similarly, Fridays for Future, inspired by the actions of young activist **Greta Thunberg**, has mobilized students worldwide in climate strikes demanding urgent policy changes. This global movement, characterized by its regular school strikes, places the spotlight on youth empowerment and the urgency of climate action. Through coordinated efforts, students have brought climate change to the forefront of public discourse, challenging governments to take decisive action. The simplicity of their message—urgent action for a sustainable future—resonates across generations, compelling leaders to respond.

The impact of these grassroots movements extends beyond protests; they play a critical role in shaping public policy and increasing climate awareness. Local governments, in response to the persistent demands of activists, have begun to adopt climate resolutions that align with the goals of these movements. These policies often include commitments to reduce carbon emissions, invest in renewable energy, and enhance community resilience against climate impacts. Moreover, the media coverage generated by these movements amplifies their message, engaging a broader audience and fostering a culture of environmental consciousness. This increased visibility fuels public engagement, encouraging individuals to participate in local initiatives and advocate for sustainable practices within their communities.

Grassroots movements thrive on their ability to mobilize and sustain efforts through innovative strategies and tools. Social media has become an indispensable ally, providing platforms for organization and outreach. Through hashtags, viral campaigns, and online petitions, movements can quickly disseminate information, rally supporters, and coordinate actions across geographic boundaries. This digital connectivity democratizes participation, allowing anyone with internet access to contribute to the cause. Additionally, community workshops and educational events serve as vital components of grassroots strategies. These gatherings provide spaces for dialogue, knowledge exchange, and skill-building, empowering participants to take action in their local contexts.

Consider the **Transition Towns initiative**, which exemplifies the power of local resilience. Originating in the UK, this grassroots movement focuses on building self-sufficient communities that prioritize sustainability and minimize their ecological footprint. By fostering local food production, renewable energy use, and waste reduction, Transition Towns creates resilient ecosystems that can withstand the pressures of global environmental change. Their success lies in their adaptability, allowing each community to tailor strategies to their unique needs and resources.

Another notable example is the **Plastic Free Communities** campaign, which tackles the pervasive issue of single-use plastics. Through grassroots advocacy, communities worldwide are implementing bans on plastic bags, straws, and other disposable items. These efforts not only reduce plastic waste but also promote a cultural shift towards sustainability. By encouraging alternatives and raising awareness about the environmental impact of plastic, these communities pave the way for broader legislative changes at regional and national levels.

As you explore the stories and achievements of grassroots movements, it's clear that collective action holds immense power in addressing climate change. These movements remind us that change often begins at the community level, where individuals unite to make a difference. Their successes inspire us to consider our role in this global effort, encouraging us to take meaningful steps toward a sustainable future.

6.2 Urban Innovations: Cities Leading the Way

Cities are at the forefront of the climate action movement, serving as hubs of innovation and leadership. Urban centers are uniquely positioned to pioneer sustainable practices and policies, given their concentration of resources, technology, and people. Many cities have committed to ambitious net-zero carbon targets, setting a precedent for environmental responsibility. This commitment is not just about reducing emissions; it is about redefining what urban living looks like in a world that prioritizes sustainability. Cities like **Copenhagen** and **Singapore** are prime examples. Their comprehensive climate action plans focus on reducing carbon footprints through a combination of policy changes, technological innovation, and community engagement. By prioritizing renewable energy, efficient public transportation, and green infrastructure, these cities are setting the stage for a sustainable future.

In the pursuit of reducing carbon emissions, urban areas are implementing innovative projects that reshape their landscapes. Green infrastructure, such as urban forests and **green roofs**, plays a vital role in this transformation. These projects not only sequester carbon but also enhance biodiversity and improve air quality. Urban forests, for example, serve as lungs for the city, absorbing CO_2 while providing shade and reducing urban heat. Singapore's "City in a Garden" initiative exemplifies this approach, integrating nature into the urban environment. Additionally, cities are developing bike-sharing programs and pedestrian zones to encourage sustainable transportation. These initiatives reduce reliance on fossil fuels, decrease traffic congestion, and promote healthier lifestyles. By reimagining transportation networks, cities are making strides toward reducing their overall carbon footprint. See "Cool Roof Project" in Chapter 8 as well.

Economic and social benefits accompany these urban innovations, contributing to growth and improving the quality of life for residents. The creation of green jobs is one such benefit, as cities invest in renewable energy projects and sustainable technologies. These jobs not only support economic development but also provide employment opportunities in emerging sectors. Furthermore, economic incentives, such as tax breaks for green businesses or subsidies for solar panel installations, stimulate local economies and encourage sustainable practices. Enhanced urban livability and public health are additional benefits of these innovations. Green spaces improve mental well-being, reduce stress, and promote physical activity, while cleaner air reduces respiratory issues and enhances overall health. These improvements make cities more attractive places to live, work, and visit.

Cities like Copenhagen serve as models for scalable solutions, offering valuable lessons for others aiming to replicate their success. Copenhagen's journey toward carbon neutrality by 2025 is a testament to the city's commitment to sustainability. Through investments in renewable energy, such as the **Middelgrunden** offshore wind farm, Copenhagen has achieved a 75% reduction in CO_2 emissions from 2005 levels. The city's extensive cycling infrastructure, spanning over 400 kilometers of dedicated bike lanes, further exemplifies its dedication to reducing emissions. Similarly, Singapore's urban sustainability initiatives, including strict green building codes and efficient public transport systems, demonstrate how policy support and technological innovation can drive climate goals. These cities emphasize the importance of community engagement, ensuring that residents are informed and involved in sustainability efforts. By fostering a culture of environmental stewardship, they inspire other cities to adopt similar strategies, paving the way for a more sustainable urban future.

6.3 Rural Resilience: Agricultural Practices for Carbon Sequestration

In the sprawling expanses of rural landscapes, a quiet revolution is taking place. Farmers and communities are adopting innovative agricultural practices that not only yield crops but also capture carbon. One of the pivotal techniques in this shift is no-till farming. Unlike conventional methods that disturb the soil, no-till farming leaves it largely intact. This practice helps in storing carbon in the soil, preventing its release into the atmosphere. It also enhances soil structure, retains moisture, and reduces erosion. As you walk through a no-till farm, you'll notice the soil teeming with life, from earthworms to microbes, all contributing to a healthier ecosystem. Yet, while no-till farming is effective, the carbon it captures is measured in megatons, a fraction when compared to the gigaton-scale reductions needed globally.

Another promising approach is **agroforestry**, which integrates trees with crops or livestock. This system not only sequesters carbon through the trees but also improves biodiversity and boosts crop yields. Imagine a landscape where fruit trees grow alongside rows of vegetables or grazing livestock. These trees act as carbon sinks, drawing CO2 from the air and storing it in their biomass and the soil. They also offer shade, reduce wind erosion, and provide habitat for wildlife. Agroforestry systems highlight how smart land management can contribute to carbon sequestration while enhancing agricultural productivity. However, like no-till farming, the scale of carbon capture remains limited to what individual plots of land can achieve and the controversy about trees only storing carbon for their life cycle of about a century. Still, the planting of a billion trees to store a gigaton of CO2, based on the standard estimate of the CDR capability of one fully grown tree, is what South Korea did after the Korean War in the 1950s.

Regenerative agriculture plays a crucial role in climate mitigation by prioritizing practices that enhance soil health and biodiversity. Techniques like cover cropping and crop rotation are central to this approach. Cover crops planted during the off-season prevent soil erosion, improve soil fertility, and capture carbon. They enrich the soil with organic matter, creating a robust environment for future crops. Meanwhile, crop rotation breaks pest and disease cycles, reducing the need for chemical inputs. The use of **biochar,** a stable form of carbon made from organic material, further boosts soil fertility and carbon storage. By incorporating biochar into the soil, farmers not only enhance crop yields but also lock away carbon for centuries. Another practice is to spread **rock dust** on farms to boost crop yields with the organic minerals that also captures CO2: https://www.newscientist.com/article/2421711-spreading-rock-dust-on-farms-boosts-crop-yields-and-captures-co2/.

Several rural initiatives exemplify the potential of these practices for significant carbon sequestration. The **Savory Institute's** holistic management practices focus on restoring grasslands through planned grazing. This approach mimics natural grazing patterns, rejuvenating the land and increasing its capacity to sequester carbon. By managing livestock movements, the Institute has improved soil health and biodiversity across millions of acres worldwide. Similarly, carbon farming programs incentivize sustainable practices by rewarding farmers for the carbon they sequester. These programs provide financial incentives for activities like tree planting, cover cropping, and reduced tillage. They demonstrate how aligning economic benefits with environmental goals can drive the widespread adoption of sustainable practices.

Despite their promise, rural carbon management faces several challenges. Access to funding and technical support remains a significant barrier for many farmers. Implementing these practices often requires upfront investments in equipment, seeds, or training, which can be prohibitive without financial assistance. Additionally, the potential for growth in sustainable agriculture depends on knowledge sharing and collaboration. By connecting farmers with researchers, policymakers, and each other, these networks can facilitate the exchange of best practices and innovations. While the current carbon capture potential of rural practices is limited to megatons, the collective impact of widespread adoption could make a meaningful contribution to broader climate goals.

6.4 Indigenous Knowledge and Practices in Carbon Management

Indigenous communities hold a profound connection with the land, one that has been cultivated through generations of wisdom and stewardship. Their traditional ecological knowledge offers invaluable insights into sustainable practices that modern science is only beginning to fully appreciate. One exemplary practice is firestick farming, a method that involves controlled burns to manage vegetation and promote ecosystem health. This technique not only reduces the risk of catastrophic wildfires but also enhances biodiversity by creating a mosaic of habitats. Controlled burns recycle nutrients, encouraging the growth of diverse plant species and creating a balanced environment where carbon sequestration can naturally occur. Such knowledge, rooted in centuries of observation and adaptation, underscores the importance of understanding and respecting ecological processes.

Traditional land stewardship extends beyond fire management, encompassing a holistic approach to conservation and resource use. Indigenous practices often emphasize the interconnectedness of all living things, fostering a deep respect for the natural world. This perspective is evident in their conservation methods, which prioritize maintaining the health of ecosystems as a whole. By managing natural resources sustainably, indigenous communities help preserve biodiversity and enhance the resilience of ecosystems to climate change. These practices contribute to carbon management by ensuring that forests, wetlands, and other natural carbon sinks remain intact and effective. In a world grappling with the impacts of climate change, the wisdom embedded in these practices offers a guiding light for sustainable living.

The integration of indigenous knowledge with modern science presents a powerful opportunity to enhance climate resilience. Collaborative research projects with indigenous communities are on the rise, as scientists recognize the value of traditional knowledge in addressing contemporary environmental challenges. These collaborations foster an exchange of ideas where scientific methods complement indigenous insights to develop innovative solutions. For example, combining traditional fire management techniques with modern fire ecology can lead to more effective fire prevention strategies. Similarly, indigenous perspectives on biodiversity can inform conservation efforts, ensuring they align with the cultural and ecological values of the land. By bridging these knowledge systems, we can create more holistic and adaptive approaches to climate resilience.

Several indigenous-led initiatives demonstrate the effectiveness of traditional practices in reducing carbon emissions and protecting biodiversity. The Indigenous Guardians program is a shining example of where indigenous communities take the lead in managing and conserving their lands. These guardians monitor ecosystems, enforce sustainable practices, and restore habitats, playing a crucial role in carbon management. Their efforts not only sequester carbon but also safeguard biodiversity, enhancing the resilience of ecosystems in the face of climate change. Such initiatives highlight the potential of indigenous leadership in environmental stewardship, offering models that can be adapted and applied in diverse contexts.

Respecting and learning from indigenous communities is not only a matter of equity but also a necessity for effective climate action. It requires acknowledging the rights and territories of indigenous peoples, ensuring they have a voice in policy-making and decision processes that affect their lands. The inclusion of indigenous voices in environmental governance enriches the dialogue, bringing diverse perspectives and solutions to the table. Protecting indigenous rights and territories is essential for preserving the cultural and ecological heritage that these communities hold. By valuing and integrating indigenous contributions, we can build partnerships that promote sustainable practices and foster climate resilience on a global scale.

As we look to the future, the lessons from indigenous knowledge remind us of the importance of harmony with nature. Their practices teach us that sustainable living is not a new concept but a timeless tradition that we must honor and integrate into our efforts to combat climate change. The collaboration between indigenous wisdom and modern science offers a path forward, one that respects the past while embracing innovation for the future.

Reflection Section: Your Role

Imagine yourself as part of a grassroots movement. How would you contribute to local environmental change? Consider organizing a community event or starting a social media campaign. Reflect on the skills and resources you can bring to the table. What steps can you take to engage others and amplify your impact?

Chapter 7

Interdisciplinary Approaches to Climate Solutions

Imagine being part of a dynamic orchestra where each instrument plays its role in a harmonious symphony, yet the conductor must ensure all elements work together seamlessly. This is the challenge of addressing climate change—an intricate problem that requires a blend of economics, policy, science, and community action. Interdisciplinary approaches offer the most promising avenues for solutions, tapping into diverse fields to combat the climate crisis with innovation and collaboration. By understanding the economic impacts, leveraging market-based incentives, and promoting sustainable models, we can create a robust framework for meaningful climate action.

The economic impacts of climate change are profound, touching every corner of the globe and influencing both local communities and global markets. **Rising temperatures,** as highlighted by recent studies, could potentially **cost the world up to 12% of its GDP for every 1°C increase,** a staggering figure that underscores the financial stakes involved. This economic strain manifests in various ways, from the devastation of natural disasters to the long-term burden of infrastructure loss. For example, hurricanes and wildfires wreak havoc on local economies, destroying homes, businesses, and public assets, while the cost of reconstruction and recovery further strains financial resources.

Moreover, dependency on fossil fuels presents a significant economic burden (see Chapter 1). As the world grapples with the transition to cleaner energy, countries heavily reliant on coal, oil, and gas face the dual challenge of restructuring their energy sectors while mitigating the associated economic risks. This dependency not only hinders progress toward emissions reduction but also ties economies to volatile global energy markets, exposing them to price fluctuations and geopolitical tensions. The shift towards renewable energy sources is an environmental imperative and an economic opportunity to break free from these constraints and foster energy independence.

Market-based incentives have emerged as powerful tools in promoting sustainable practices and driving the adoption of green technologies across industries. Carbon pricing mechanisms, such as carbon taxes and cap-and-trade systems, assign a cost to carbon emissions, encouraging companies to reduce their environmental footprint. By internalizing the social cost of carbon, these mechanisms create financial incentives for businesses to invest in cleaner technologies and improve efficiency. Tax credits for renewable energy investments further stimulate the market, providing financial benefits to individuals and corporations that embrace solar, wind, and other sustainable energy sources.

Evaluating the effectiveness of economic policies in reducing emissions reveals promising outcomes. **Cap-and-trade systems**, for instance, have proven successful in regions like California, where emissions have consistently fallen while the economy has grown. These systems set a cap on total emissions and allow companies to trade allowances, fostering a competitive market that rewards innovation and reduction efforts. Similarly, subsidies for electric vehicles and public transport have accelerated the transition to cleaner transportation options, reducing reliance on fossil fuels and lowering urban pollution levels.

The concept of a **green economy** represents a transformative shift towards sustainable economic models that prioritize environmental health. Embracing circular economy principles, where waste is minimized and resources are reused, can significantly reduce environmental impact while boosting economic resilience. This approach not only conserves natural resources but also creates new business opportunities in recycling, remanufacturing, and sustainable product design. The growth of green finance and investment further supports this transition, directing capital towards projects and companies that align with environmental and social governance criteria.

7.1 Visual Element: Economics of Climate Change Infographic

Imagine an infographic illustrating the interconnectedness of climate change and the economy. It should highlight key statistics such as GDP losses per degree of warming, the financial burden of fossil fuel dependency, and the impact of market-based incentives. This visual representation can underscore the economic rationale for pursuing sustainable practices and policies, such as the one below from Statista.com/chart/11673 and the World Meteorological Organization that reaches to about **$1.5 trillion cost by 2019**. We can only guess what this decade will bring.

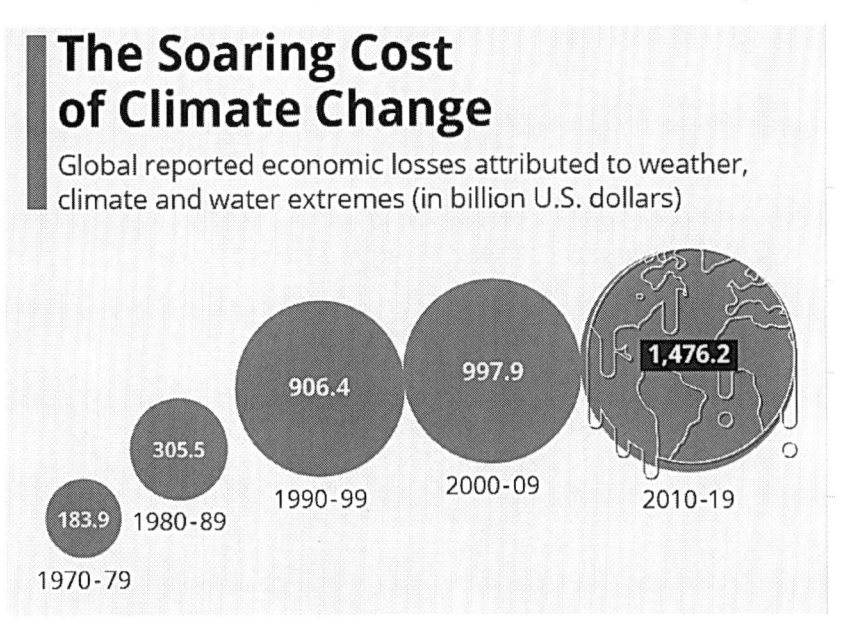

The Soaring Cost of Climate Change

Global reported economic losses attributed to weather, climate and water extremes (in billion U.S. dollars)

- 183.9 — 1970-79
- 305.5 — 1980-89
- 906.4 — 1990-99
- 997.9 — 2000-09
- 1,476.2 — 2010-19

As we navigate the economic dimensions of climate change, the need for interdisciplinary approaches becomes evident. By integrating economic strategies with environmental goals, we can create a more resilient and sustainable future. The collaboration between policymakers, businesses, and communities will be instrumental in driving the changes necessary to address the climate crisis effectively.

7.2 Policy Synergies: Aligning National and Local Efforts

Imagine a sprawling landscape of policies, each crafted with well-meaning intentions but often existing in isolated silos, missing the opportunity for greater impact through synergy. This is the reality of climate policy when national goals and local initiatives operate independently. The essence of effective climate action lies in the seamless integration of policies across all levels of governance. When national climate goals align with local initiatives, they create a unified front that is far more effective than the sum of its parts. This alignment ensures that efforts at every level are coordinated, avoiding duplication and maximizing resources.

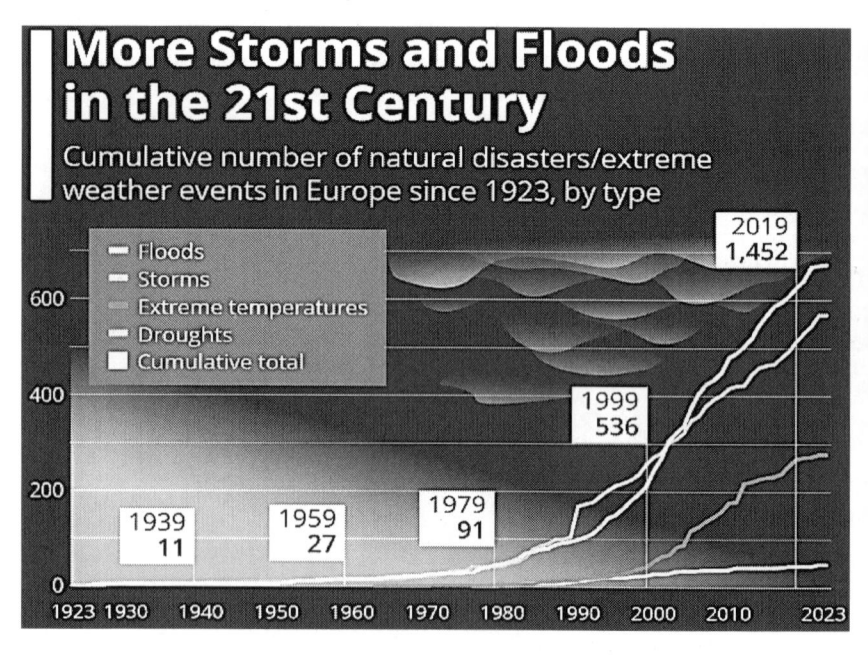

More Storms and Floods in the 21st Century

Cumulative number of natural disasters/extreme weather events in Europe since 1923, by type

- Floods
- Storms
- Extreme temperatures
- Droughts
- Cumulative total

2019: 1,452
1999: 536
1979: 91
1959: 27
1939: 11

Source: International Disasters Database (EM-DAT)

National policies provide the framework and resources, while local governments adapt these strategies to the unique needs and circumstances of their communities, ensuring relevance and effectiveness. A carbon tax with revenue gained going toward renewable energy development is a proposed policy change that has some traction in government circles.

Germany's **Energiewende** exemplifies the power of policy alignment. This ambitious energy transition initiative aims to shift from fossil fuels to renewable energy, reducing greenhouse gas emissions and enhancing energy efficiency. At the national level, comprehensive policies were set to drive this transition, including targets for renewable energy production and energy-saving measures. These goals were not imposed top-down but integrated with regional plans, allowing local governments to tailor their approaches. Regions like Bavaria and North Rhine-Westphalia developed their own energy strategies, focusing on local strengths such as wind and solar power. This synergy between national and local efforts has propelled Germany to become a leader in renewable energy, showing how coordinated policies can drive significant climate progress.

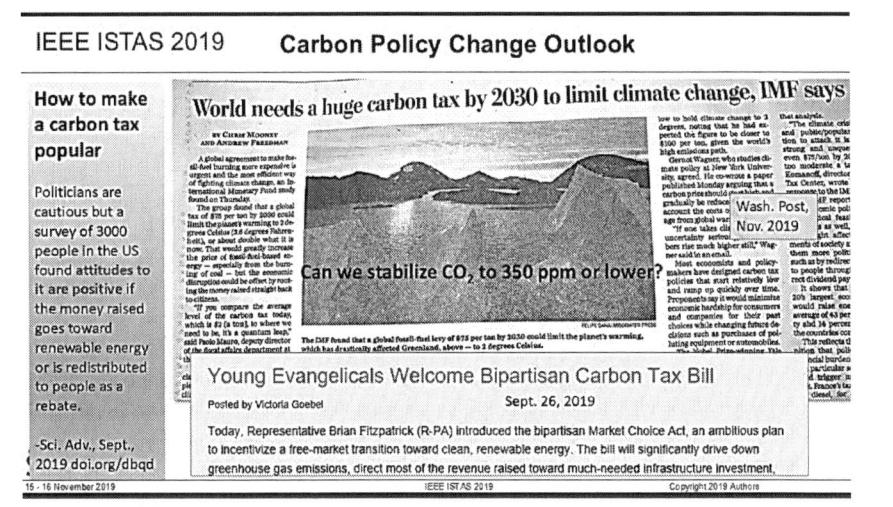

IEEE ISTAS 2019 — Carbon Policy Change Outlook

How to make a carbon tax popular
Politicians are cautious but a survey of 3000 people in the US found attitudes to it are positive if the money raised goes toward renewable energy or is redistributed to people as a rebate.
-Sci. Adv., Sept., 2019 doi.org/dbqd

World needs a huge carbon tax by 2030 to limit climate change, IMF says
Wash. Post, Nov. 2019
Can we stabilize CO₂ to 350 ppm or lower?

Young Evangelicals Welcome Bipartisan Carbon Tax Bill
Posted by Victoria Goebel — Sept. 26, 2019
Today, Representative Brian Fitzpatrick (R-PA) introduced the bipartisan Market Choice Act, an ambitious plan to incentivize a free-market transition toward clean, renewable energy. The bill will significantly drive down greenhouse gas emissions, direct most of the revenue raised toward much-needed infrastructure investment,

Similarly, **California** provides a model for state and federal collaboration in climate policy. Known for its stringent environmental regulations, California has implemented policies that align with and often exceed federal standards. The state's cap-and-trade program, which sets a limit on emissions and allows for trading of allowances, complements federal efforts to reduce greenhouse gases. Additionally, **California's Renewable Portfolio Standard** mandates a specific percentage of electricity to come from renewable sources, pushing utilities to adopt cleaner energy. These policies align with national goals, creating a cohesive framework that enhances the state's ability to meet and exceed emissions targets.

Despite these successes, policy fragmentation remains a significant challenge. Divergent priorities, political agendas, and resource allocations can hinder cohesive action. Overcoming these obstacles requires robust inter-agency coordination mechanisms that facilitate communication and collaboration among various governmental bodies. Establishing clear channels for dialogue and decision-making ensures that policies are aligned and that all stakeholders are working toward common objectives. Stakeholder engagement processes are equally vital, bringing together government officials, industry leaders, community representatives, and environmental advocates to share insights and address concerns. By fostering an inclusive environment, these processes build consensus and support for integrated climate policies.

Effective policy integration relies on practical tools and strategies that bridge gaps between different governance levels. Policy harmonization tools, such as standardized guidelines and frameworks, help align objectives and metrics, ensuring consistency across regions. These tools enable policymakers to assess progress and make informed adjustments to strategies. Joint task forces and intergovernmental panels play a crucial role in facilitating collaboration, bringing together representatives from federal, state, and local governments to coordinate efforts. These groups serve as platforms for sharing best practices, identifying synergies, and addressing challenges collectively. By working together, policymakers can leverage the strengths of each level of government, creating a more resilient and effective approach to climate action.

7.3 Sociological Insights into Climate Behavior Change

Understanding the power of societal norms and values is crucial in shaping climate action. These cultural narratives often guide our behavior, like an invisible hand pushing us toward sustainability—or not. Think about how recycling has become second nature in many communities. This shift didn't happen overnight; it was driven by changing perceptions about waste and responsibility. As more people adopted recycling, it became a social norm, illustrating how cultural values can influence environmental behaviors. Social movements play a pivotal role in this transformation. They act as catalysts for change, challenging existing norms and advocating for new ones that prioritize the planet's health. Movements like Fridays for Future have leveraged cultural narratives to promote sustainability, capturing the public's imagination and urging action. By reshaping societal values, they create a fertile ground for pro-environmental behaviors to take root and flourish.

Promoting sustainable practices within communities requires strategic approaches that resonate with people's everyday lives. Community-based social marketing is one such strategy, focusing on encouraging positive behaviors by understanding and addressing barriers to change. This approach often involves direct engagement with community members, identifying the specific challenges they face, and tailoring solutions to meet those needs. Techniques like setting up local recycling programs or organizing community clean-up events can foster a sense of involvement and ownership, motivating individuals to adopt sustainable habits. For instance, placing recycling bins in convenient locations or providing incentives for using public transport can nudge individuals toward more sustainable actions. These techniques rely on the principle that small changes in the environment can lead to significant shifts in behavior, creating a ripple effect of positive environmental impact.

Education and awareness campaigns are powerful drivers of behavior change, shaping public perception and influencing actions. School programs focused on climate literacy serve as foundational tools, equipping students with the knowledge and skills needed to understand and address environmental challenges. By integrating climate education into the curriculum, schools can foster a generation of environmentally conscious citizens who are prepared to tackle climate issues with informed perspectives. Media campaigns also play a crucial role in raising awareness about climate change, reaching broad audiences through various channels. Effective campaigns often use compelling storytelling and visuals to convey their message, making complex issues accessible and engaging. By highlighting the urgency of climate action and presenting practical solutions, these campaigns can inspire individuals to make more sustainable choices in their daily lives.

Examining successful behavior change initiatives reveals valuable insights into what works in fostering societal shifts. The implementation of plastic bag bans in cities around the world provides a compelling example. By prohibiting single-use plastic bags, these bans encourage people to switch to reusable alternatives, leading to a significant decrease in plastic waste. This policy change, coupled with public education efforts, has resulted in widespread behavioral shifts, illustrating the power of targeted interventions. Another example is the introduction of incentives for public transport use, which has led to increased ridership and reduced reliance on private vehicles. By offering discounted fares or enhanced services, these initiatives make public transport a more attractive option, encouraging people to leave their cars behind. These case studies demonstrate that strategic interventions, supported by education and incentives, can lead to meaningful changes in societal behaviors, fostering a culture of sustainability.

7.4 Collaborative Frameworks: Bridging Science and Policy

In the intricate dance of climate change mitigation, collaboration between scientists and policymakers is vital. Imagine scientists as the composers of a symphony, crafting the melodies of understanding through rigorous research and data. Policymakers, then, are the conductors, translating these complex notes into actionable policies that guide society. This delicate interplay requires scientific advisory panels to serve as bridges, ensuring that the latest research informs policy decisions. These panels consist of experts who provide guidance on issues ranging from carbon emissions to renewable energy. Their role is crucial, as they distill complex scientific data into practical recommendations that policymakers can implement. By offering a clear line of communication between science and policy, these panels help ensure that decisions are grounded in the most current and comprehensive knowledge available.

Case studies highlight the transformative power of science-driven policy reforms. Consider the **Montreal Protocol**, an agreement aimed at phasing out substances that deplete the ozone layer. This landmark treaty was successful because it was rooted in scientific understanding, with policymakers working closely with scientists to develop effective solutions. The result has been a significant reduction in the use of harmful chemicals and a gradual recovery of the ozone layer. Another example is **the Intergovernmental Panel on Climate Change (IPCC)**, which synthesizes scientific research to inform global climate agreements. Its reports have played a pivotal role in shaping international climate policies, demonstrating how science can lead to impactful change when integrated into the policy-making process.

However, it is not well known that **oil companies have executives serving on the IPCC Board.** Furthermore, it was recently publicly disclosed by an IPCC Board Member that those executives have a large influence on the **Summary Section** of the IPCC Reports, which is what the majority of the public actually reads. So, for example, when the IPCC Board votes to state that "we are hopeful that the world can keep the global temperature rise below a 1.5 or 2°C level," the public actually believes it, though the physics of the amount of CO_2 accumulation above our heads surpasses such a small limit to global warming. Such obfuscation amounts to manipulation of the facts, while using the public's gullibility to foster fossil fuel usage.

To facilitate science-policy integration, various frameworks have emerged, promoting collaboration and shared knowledge. One such model is the co-production of knowledge, where scientists and policymakers work together from the outset to define research questions and design studies. This collaborative approach ensures that research outcomes are directly relevant to policy needs, enhancing their applicability and impact. Interdisciplinary research networks also play a key role, bringing together experts from diverse fields to tackle complex climate challenges. These networks foster innovation by encouraging cross-pollination of ideas and methodologies, leading to more holistic solutions.

Despite these promising models, bridging the gap between science and policy presents challenges. Communication gaps often arise due to differing priorities and languages used by scientists and policymakers. Scientists may focus on precision and nuance, while policymakers require clear, actionable insights. These differences can hinder effective collaboration, as each group may struggle to understand the other's perspective. To overcome this, joint research initiatives can provide a platform for dialogue, allowing stakeholders to align their goals and expectations. By fostering mutual understanding, these initiatives can enhance the integration of scientific knowledge into policy frameworks.

Successful examples of science-policy collaboration offer valuable lessons for future efforts. The **Montreal Protocol's** success in addressing ozone depletion demonstrates the power of science-informed policy action. By leveraging scientific research, policymakers were able to develop targeted regulations that have had a lasting positive impact on the environment. Similarly, the IPCC's influence on global climate agreements underscores the importance of synthesizing scientific knowledge to guide international action.

These case studies highlight the potential for meaningful change when science and policy work in tandem, underscoring the need for continued collaboration to address the pressing challenges of climate change.

Reflection Section

As we reflect on the necessity of collaborative frameworks, how do we integrate scientific knowledge into policy decisions, in today's political climate, when it is crucial for effective climate action. How can we foster partnerships between scientists and policymakers to ensure that decisions are informed by the best available evidence, leading to more effective and sustainable outcomes? This integration is not just beneficial; it is essential for navigating the complexities of climate change and creating a more resilient future.

Chapter 8

Practical Steps for Individuals

Imagine waking up each morning knowing that your daily decisions can contribute to a healthier planet. The lowering of CO2 emissions must go in parallel with the removal of CO2 from the atmosphere. This empowerment starts with understanding your carbon footprint—a measure of the total greenhouse gases you produce through everyday activities. It's not just about the big gestures; every small action counts. By examining how much carbon dioxide you emit, you can make informed choices that lead to significant change. Think of it as a personal audit, allowing you to see where your lifestyle impacts the environment most. This self-awareness is the first step toward meaningful action, enabling you to reduce your emissions and inspire others to do the same.

8.1 Carbon Footprint Calculators: Know Your Impact

Carbon footprint calculators are invaluable tools that help you take stock of your environmental impact. These online platforms, like the EPA's Household Carbon Footprint Calculator and the CoolClimate Network, provide an accessible way to quantify the emissions resulting from your daily activities, such as using electricity, commuting, and generating waste. By inputting specific data related to your lifestyle, these calculators offer a detailed analysis of your carbon footprint, highlighting areas where you can make the most effective changes. They serve as a starting point for understanding the scale of your emissions and setting goals for reduction, making sustainability more approachable and actionable.

Using a carbon footprint calculator begins with gathering information about your energy consumption, transportation habits, and lifestyle choices. Start by recording your monthly electricity and gas usage, which often involves checking your utility bills. Input this information into the calculator, along with details about your transportation habits, such as the type of vehicle you drive and the number of miles you travel per week. Additional lifestyle factors, like dietary preferences and waste generation, also contribute to your total emissions. Many calculators offer the option to use default values based on national averages, but for the most accurate results, it's best to use your own data. This detailed input process ensures that the calculator provides a precise estimate of your carbon footprint, offering insights into areas that require attention and improvement.

Understanding your carbon footprint is more than just a numerical exercise; it is a powerful motivator for change. Awareness of your emissions can drive targeted actions to reduce your carbon output. By pinpointing the major sources of emissions in your life, such as high energy consumption or frequent car travel, you can prioritize changes that will have the most significant impact. For example, if the calculator shows that your home energy use is a major contributor, you might consider investing in energy-efficient appliances or improving your home's insulation. Setting realistic reduction goals based on this knowledge helps you track progress and stay motivated, as you can see the tangible benefits of your efforts over time.

Once you have identified the key contributors to your carbon footprint, you can explore specific strategies to reduce your emissions. One effective approach is to reduce energy use at home. This can involve simple actions like turning off lights when they're not needed, using programmable thermostats, or switching to LED bulbs. For those who drive frequently, opting for sustainable transportation methods can make a significant difference. Consider carpooling, using public transit, or biking whenever possible. These changes not only reduce emissions but can also lead to cost savings and improved health. Additionally, exploring renewable energy options, like solar panels, can further decrease your reliance on fossil fuels and contribute to a cleaner environment.

> **Cool Roof and Attic Concepts** - Reducing carbon emissions is also possible by using and advocating the Cool Roof Concept: Primarily, **White Roofs** absorb considerably less heat. Secondly, **implementing Phase Change Materials** for insulation, *instead of fiberglass* for example, saves a huge amount both in heating and in air conditioning.

The president of InsolCorp told me at an AEE trade show that he was able to install their **Infinite R** phase change insulation along the inside surface of a company's production warehouse roof so that they did not need to invest in air conditioning at all! The slide on this page explains the amazing outdoor swing in temperature while the Infinite R insulated inside temperature oscillates only a couple of degrees. Now if your personal advocacy and solicitation can convince InsolCorp to start distributing their superior phase change insulation to Home Depot and Ace Hardware stores for the public, then we could be using it our attics too. They seem to cater only to large company orders so far.

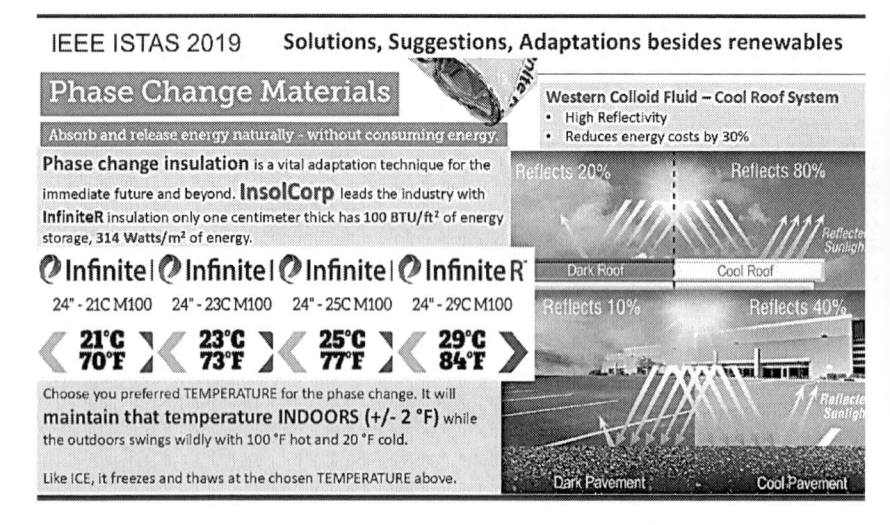

IEEE ISTAS 2019   Solutions, Suggestions, Adaptations besides renewables

Phase Change Materials
Absorb and release energy naturally - without consuming energy
Phase change insulation is a vital adaptation technique for the immediate future and beyond. **InsolCorp** leads the industry with **InfiniteR** insulation only one centimeter thick has 100 BTU/ft² of energy storage, 314 Watts/m² of energy.

Ⓘ Infinite I Ⓘ Infinite I Ⓘ Infinite I Ⓘ Infinite R
24" - 21C M100   24" - 23C M100   24" - 25C M100   24" - 29C M100

21°C / 70°F   **23°C / 73°F**   **25°C / 77°F**   **29°C / 84°F**

Choose you preferred TEMPERATURE for the phase change. It will **maintain that temperature INDOORS (+/- 2 °F)** while the outdoors swings wildly with 100 °F hot and 20 °F cold.
Like ICE, it freezes and thaws at the chosen TEMPERATURE above.

Western Colloid Fluid – Cool Roof System
- High Reflectivity
- Reduces energy costs by 30%

Reflects 20%   Reflects 80%
Dark Roof   Cool Roof
Reflects 10%   Reflects 40%
Dark Pavement   Cool Pavement

8.2 Interactive Element: Personal Carbon Reduction Plan

Create a personalized carbon reduction plan by listing three specific actions you can take to lower your emissions. Use the results from your carbon footprint calculator to guide your choices, focusing on areas where you can make the biggest impact. Set a timeline for implementation and track your progress over the next few months.

By taking these practical steps, you contribute to a collective effort in combating climate change. While individual actions may seem small in isolation, they add up to create a significant impact. Your commitment to reducing your carbon footprint not only benefits the environment but also sets an example for others, inspiring widespread change and fostering a culture of sustainability.

8.3 Sustainable Living: Everyday Actions with Major Impact

Every day, you make choices that can contribute to a healthier environment. These decisions, though sometimes small, add up to create a significant impact on your carbon footprint. Consider switching to energy-efficient appliances. These devices consume less power and often perform better, saving you money on energy bills while reducing emissions. Simple changes, like using LED light bulbs or choosing appliances with high energy ratings, can lead to substantial energy savings over time. Another effective measure is reducing water usage. By adopting mindful habits such as fixing leaks, taking shorter showers, and using water-efficient fixtures, you can conserve this precious resource and lower the energy needed for water heating.

Sustainable consumption practices extend beyond energy and water use. Every product you purchase carries an environmental footprint, from the resources used in production to the waste it generates. You can minimize this impact by choosing products with minimal packaging. Opt for items that use recycled materials or are packaged in biodegradable containers. By supporting local and seasonal produce, you reduce the carbon emissions associated with transporting goods over long distances. This not only supports your local economy but also ensures fresher and often more nutritious food. Additionally, using reusable bags, bottles, and containers reduces your reliance on single-use plastics, a major contributor to environmental pollution.

Technology plays a pivotal role in facilitating sustainable living. Various apps and tools can guide you toward more eco-friendly choices. Eco-friendly shopping apps, such as Good On You, offer insights into the sustainability practices of fashion brands, empowering you to make informed decisions. Energy monitoring devices for home use provide real-time data on your energy consumption, helping you identify areas where you can reduce usage. These tools make it easier to track your progress and stay motivated on your sustainability journey. By integrating technology into your daily routine, you can make more informed decisions that align with your environmental values.

Success stories abound, showcasing individuals and communities that have significantly reduced their carbon footprint. Consider those living a zero-waste lifestyle, where every effort is made to eliminate waste by composting, recycling, and reusing materials. These advocates demonstrate that it is possible to live sustainably without sacrificing convenience or quality of life. Similarly, families achieving net-zero energy homes have transformed their living spaces into models of efficiency and sustainability. By incorporating renewable energy sources, enhancing insulation, and optimizing energy use, they have reduced their reliance on external power sources. These examples serve as powerful motivators, illustrating the tangible benefits of sustainable living.

Cities worldwide are also making strides toward sustainability, committing to ambitious goals of reaching net-zero energy. San Francisco, Boston, and Austin are leading the charge, implementing policies and initiatives to reduce emissions and promote renewable energy. These cities serve as blueprints for others, showcasing the potential for urban centers to drive meaningful change. Meanwhile, cities like Copenhagen, Johannesburg, and Montreal are committed to making buildings net-zero, setting standards for energy efficiency and sustainability. By renovating existing structures and designing new ones with green principles, these cities are creating healthier environments for their residents while contributing to the global fight against climate change.

8.4 Engaging in Local Advocacy: From Awareness to Action

Engaging in local advocacy for environmental change is not just about personal commitment—it's about fostering community action that amplifies your efforts. Imagine being part of a vibrant network where your voice, combined with others, can drive significant change. Joining local environmental groups or clubs is a practical first step. These groups provide a platform for sharing ideas, organizing events, and mobilizing resources. They offer a sense of community and shared purpose, making the daunting task of combating climate change more approachable. By participating in these groups, you connect with like-minded individuals who are equally passionate about the environment, allowing you to exchange knowledge and strategies that can lead to impactful action.

One effective way to advocate for sustainable practices is by organizing community clean-up events. These gatherings not only beautify local areas but also raise awareness about the importance of environmental stewardship. They serve as a visible testament to the power of collective action, inspiring others to join the cause. Additionally, lobbying for local government policy changes can create a lasting impact. This involves attending town meetings, writing letters, and presenting petitions to local officials. By advocating for policies that promote sustainability, such as stricter recycling regulations or the development of green spaces, you can influence decisions that shape your community's environmental future. Your involvement in these activities demonstrates a commitment to creating a healthier, more sustainable world for future generations.

Building local networks further amplifies your advocacy efforts. Creating or joining neighborhood sustainability committees allows you to collaborate with others in developing initiatives tailored to your community's needs. These committees can focus on projects like community gardens, renewable energy adoption, or water conservation programs. Networking with local businesses and schools can also be incredibly effective. Businesses can be encouraged to adopt sustainable practices, such as reducing waste or sourcing eco-friendly materials, while schools can integrate environmental education into their curricula. These partnerships create a holistic approach to sustainability, engaging all aspects of the community in meaningful action. By fostering these relationships, you build a robust network that supports and expands your advocacy efforts.

In today's digital age, social media plays a crucial role in environmental advocacy. These platforms offer a powerful tool for raising awareness and mobilizing action on a scale previously unimaginable. Creating impactful social media campaigns allows you to reach a broader audience, sharing information about environmental issues and encouraging others to get involved. You can use these platforms to promote events, share educational resources, and highlight success stories that inspire action. By leveraging the reach of social media, you can engage individuals who may not be directly involved in environmental groups, broadening the impact of your advocacy. Sharing success stories online not only celebrates achievements but also provides a model for others to replicate, further spreading the message of sustainability.

8.5 Empowering Education: Teaching Climate Literacy at Home and in School

Understanding climate science is more than a classroom endeavor; it is a crucial skill for making informed decisions in today's world. Climate literacy involves comprehending the complex interactions between the Earth's systems and recognizing the impact of human activities on these systems. It encompasses understanding scientific principles, grasping the social implications, and being aware of the policy frameworks that guide climate action. For high school and college students, this knowledge lays the groundwork for conscious decision-making and responsible citizenship. Yet, climate literacy should not be confined to academic settings; it starts at home, where conversations can lead to a deeper understanding and lifelong commitment to sustainability.

Integrating climate education into everyday family life need not be daunting. Begin by discussing climate topics during family meals or outings, where informal settings encourage open dialogue. Share interesting facts or recent news stories about climate change, sparking curiosity and discussion. You can also incorporate educational materials into home learning. Use documentaries or online courses to explore topics like renewable energy or biodiversity. Platforms like **NASA's Climate Kids** offer engaging, age-appropriate content that simplifies complex topics, making them accessible to young learners. This approach not only increases understanding but also fosters a culture of inquiry and critical thinking. By making climate education a family affair, you reinforce its importance and relevance to everyday life.

Several resources can aid in teaching climate concepts at home. Books on environmental themes, such as the textbook I used when teaching Environmental Science at Erie Community College in Buffalo NY: "*Living in the Environment*" by G. Tyler Miller, provide comprehensive insights into ecological principles and sustainability issues. Documentaries like "Our Planet" on YouTube showcase the beauty and fragility of the natural world, illustrating the impacts of climate change in a visually compelling way. Online courses and educational websites offer structured learning experiences that delve into specific aspects of climate science. These resources present information in a relatable manner, bridging the gap between academic jargon and everyday understanding. By leveraging these tools, you equip yourself and your family with the knowledge necessary to engage in informed discussions and make conscious choices that benefit the planet.

Fostering climate literacy at home has far-reaching implications. Educated individuals are better equipped to influence broader societal change. They develop critical thinking skills that enable them to analyze information, question assumptions, and propose solutions. This mindset encourages proactive behaviors, such as advocating for sustainable practices or supporting policies that mitigate climate change. Moreover, by instilling these values in younger generations, you inspire them to become advocates for sustainability. They gain the confidence to voice their opinions, participate in civic activities, and lead initiatives that promote environmental stewardship. This ripple effect extends beyond the home, contributing to a more informed and engaged society.

Empowering education in climate science is a powerful step toward building a sustainable future. It starts with understanding the basics, engaging in conversations, and utilizing available resources to deepen knowledge. The impact of fostering climate literacy at home is profound, creating individuals who are not only aware of the challenges but also motivated to seek solutions. As you nurture this understanding within your family, you contribute to a collective effort to address climate change, empowering the next generation to lead with knowledge and conviction.

Reflection Section: Local Advocacy Tools

Engaging in local advocacy is a dynamic and rewarding process. It connects you with your community, empowers you to effect change, and creates a ripple effect that can lead to broader societal impact. Your efforts, combined with those of others, can drive meaningful progress in the fight against climate change. As you participate in these initiatives, you become part of a larger movement, one that is working tirelessly to protect and preserve our planet for generations to come.

Consider compiling a list of tools and resources that can aid in your advocacy efforts. This might include contact information for local environmental groups, guides for organizing community events, and templates for reaching out to local officials. Having these resources at your fingertips can streamline your efforts and empower you to take action confidently, such as with a presentation of one of this author's climate videos to a local group. Showing the YouTube video "Six Degrees Could Change the World" may be just the one for a nontechnical crowd to motivate them.

Adding a few factoids to your advocacy activity can always help, such as from Chapter 1:

1) Those who think renewables will solve climate change don't realize that **Global Energy Usage** is 90% fossil fuel based;
2) CO2 Emissions were a tolerable **10 gigatons** (10 billion tons) per year in 1960. But every 20 years since, humans have added an equal amount, compounding the storage bin of heat-trapping gas above us, until after 2020, we emit about **40 Gt/yr** which is only the latest number in the upward sloping graph of CO2 global Carbon Dioxide Emissions in Chapter 1.
3) The earth is indebted for about +6°C from the excess Teraton of CO2, according to the most accurate and on target climatologist, Jim Hansen. Experts agree that we are realizing about +2°C right now out of that 6°C and even Stanford U says we probably will realize the full +6°C by 2100, which will feel like a threshold of an impending 22nd century extinction.
4) The most prophetic and disturbing factoid adds +0.36°C/decade or about 1 to 2 ppm every year, from the latest Hansen 2025 finding, so that our world seems determined to add another +1°C to our 6°C indebtedness every 1 to 3 decades now and every decade after 2050, so we will be repeating **the PETM** of a +10°C to +12°C realized indebtedness as we enter the actual, widespread 22nd century extinction.
5) These four factoids should allow the trillion dollar estimate for gigaton CDR to be a welcome conclusion to this impending doom scenario for any audience hopefully. It really is an investment without any requirement for recurring expenditure (unlike prescription drugs), if done with renewable energy and durable hardware.

Chapter 9

Global Perspectives and Cooperation

Imagine a world where nations unite to tackle climate challenges, transcending borders and differences to forge a path toward a sustainable future. This vision is at the heart of the Paris Agreement, a groundbreaking accord that represents a collective commitment to limit global warming. Signed by nearly every country on Earth, this landmark treaty aims to keep the temperature rise well below 2°C, striving for a more ambitious goal of 1.5°C above pre-industrial levels. The Paris Agreement emerged from the realization that climate change knows no boundaries, and its impacts will affect everyone. Focusing on renewable energy solutions laid the groundwork for a global transition toward sustainable practices. However, it fell short by not adequately addressing gigaton-level carbon dioxide removal (CDR) technologies, which are crucial for reversing the current warming trend.

The **Paris Agreement** introduced **Nationally Determined Contributions (NDCs)**, where each country pledges its emission reduction targets. These NDCs are vital for tracking progress and ensuring accountability. Yet, as of today, many signatories have struggled to meet their commitments. Countries have made varying levels of progress, with some, like the European Union, taking significant strides in reducing emissions, while others lag behind. The **UN Environment Programme's Emissions Gap Report** highlights the gap between current pledges and the necessary reductions to avoid catastrophic warming. Current pledges might lead to a temperature rise of 2.5-2.9°C, far exceeding the Paris goals. This discrepancy underscores the need for nations to revisit their NDCs and set more ambitious targets that align with the urgency of the climate crisis.

Since the Paris Agreement, the global community has witnessed both progress and setbacks in achieving its goals. On one hand, renewable energy adoption has accelerated, with solar and wind power becoming more cost-competitive. On the other hand, global emissions continue to rise, reaching record levels in 2022. Developing nations face unique challenges in meeting their NDCs, often grappling with financial constraints and inadequate infrastructure. These barriers highlight the importance of international support and collaboration, ensuring that all countries have the resources needed to transition to a low-carbon economy. Reports on country compliance reveal a mixed picture, with some nations making impressive strides while others fall short. The challenges are complex and require a concerted effort from all stakeholders to overcome.

In the aftermath of the Paris Agreement, new initiatives and actions have emerged to build on its foundation. The Katowice Climate Package, adopted in 2018, provided a rulebook for implementing the Paris Agreement, emphasizing transparency and accountability. This package laid the groundwork for monitoring and reporting emissions reductions, ensuring that countries stay on track. Stanford University's "**100% in 139 Countries**" initiative aims to support nations in achieving carbon neutrality, offering a roadmap for transitioning to sustainable energy systems.

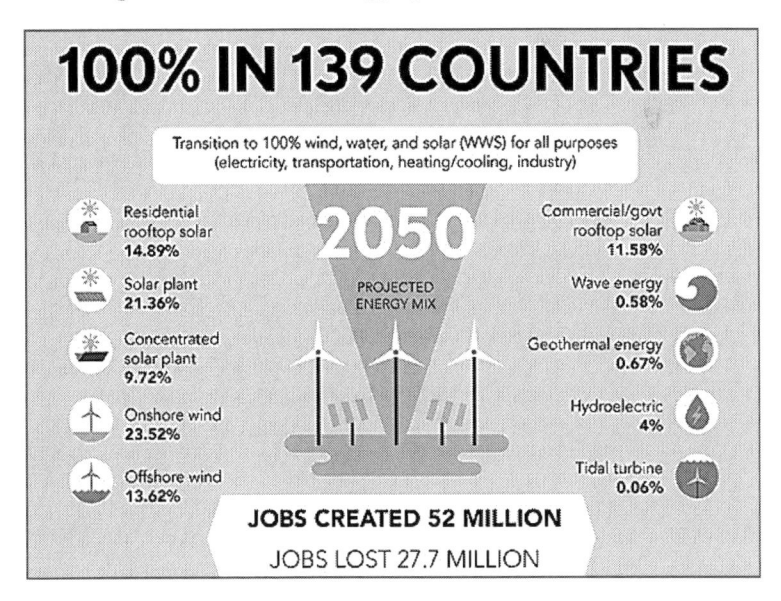

100% IN 139 COUNTRIES

Transition to 100% wind, water, and solar (WWS) for all purposes
(electricity, transportation, heating/cooling, industry)

2050
PROJECTED ENERGY MIX

- Residential rooftop solar **14.89%**
- Solar plant **21.36%**
- Concentrated solar plant **9.72%**
- Onshore wind **23.52%**
- Offshore wind **13.62%**
- Commercial/govt rooftop solar **11.58%**
- Wave energy **0.58%**
- Geothermal energy **0.67%**
- Hydroelectric **4%**
- Tidal turbine **0.06%**

JOBS CREATED 52 MILLION
JOBS LOST 27.7 MILLION

Dr. Mark Jacobson directs **Stanford's Atmosphere and Energy Program** and authored this initiative for a transition to 100% wind, water, and solar for all purposes (electricity, transportation, heating/cooling, industry) by 2050 for 139 countries ("100% Clean and Renewable Wind, Water, and Sunlight All-Sector Energy Roadmaps for 139 Countries of the World" – Joule, V.1, N.1, p. 108, 9/6,17). His short YouTube video also explains the process nicely.

Additionally, the **Green New Deal Implementation Guide** by Senator Markey, released in July 2024, outlines a comprehensive strategy for the U.S. to achieve net-zero emissions by mid-century. These post-Paris initiatives reflect a growing recognition of the need for bold action and innovative solutions to address the climate crisis.

9.1 Global Cooperation and Interaction

International diplomacy plays a crucial role in shaping global climate policy and facilitating cooperation and coordination among nations. The United Nations Framework Convention on Climate Change (UNFCCC) hosts annual conferences, known as COPs, where countries negotiate and update their commitments. These conferences provide a platform for dialogue, enabling nations to share experiences, challenges, and best practices. Climate envoys and negotiators act as key intermediaries, forging agreements and building consensus on critical issues. COP 29 will be another pivotal moment, offering an opportunity to increase climate ambition and set more rigorous targets for future NDCs. As the world grapples with the realities of climate change, diplomatic efforts will continue to be a cornerstone of global climate action, fostering collaboration and unity in pursuit of a sustainable future.

9.2 Cross-Border Carbon Trading and Its Impacts

Imagine a bustling global marketplace, not for goods or services, but for carbon emissions. This is the world of carbon trading, where countries and companies buy and sell emissions allowances to meet environmental targets. At the heart of this system lies the concept of cap-and-trade, a mechanism designed to incentivize reductions in greenhouse gas emissions. Under cap-and-trade, a government sets a cap on the total amount of emissions allowed, issuing permits that represent the right to emit a specific amount of CO_2. Companies that reduce their emissions below their allotted amount can sell their excess permits to others, creating a financial incentive to cut emissions. This market-driven approach encourages innovation and efficiency as businesses seek cost-effective ways to reduce their carbon footprint.

Alongside cap-and-trade, carbon offset credits provide another tool for emissions reduction. These credits represent a reduction in emissions achieved by projects outside the regulated sector, such as reforestation or renewable energy initiatives. Companies can purchase these offsets to compensate for their emissions, supporting sustainable projects while meeting regulatory requirements. The integration of offset credits into carbon markets expands the scope of emissions reductions, allowing for a more comprehensive approach to tackling climate change. However, the effectiveness of offsets depends on rigorous standards and verification processes to ensure that the reductions are real and additional. Without these safeguards, the risk of greenwashing—where businesses make misleading claims about their environmental impact—looms large.

Carbon trading offers several benefits, primarily by providing economic incentives for emissions reductions. By putting a price on carbon, these systems encourage businesses to invest in cleaner technologies and practices. The flexibility of trading allows companies to tailor their strategies to their specific circumstances, driving innovation and efficiency. Moreover, carbon markets can generate significant revenue for governments, which can be reinvested in climate initiatives and social programs. Yet, the system is not without its drawbacks. Market manipulation poses a significant risk, as entities may engage in speculative trading or hoarding of permits to influence prices. These practices can undermine the integrity of the market, reducing its effectiveness in achieving emissions reductions. Additionally, the complexity of carbon markets can create barriers for smaller businesses, limiting their participation and benefits from the system.

Successful examples of international carbon markets demonstrate the potential of cross-border trading systems. The **European Union Emissions Trading System (EU ETS)** is the largest and most established carbon market, covering over 11,000 installations across Europe. Since its inception in 2005, the EU ETS has driven significant emissions reductions in sectors like power and industry, demonstrating the effectiveness of cap-and-trade at scale. Similarly, the **California-Quebec carbon market** linkage represents a successful collaboration between two regions, creating a larger and more liquid market. This cross-border initiative has harmonized emissions reduction efforts, allowing for greater flexibility and cost-effectiveness. By pooling resources and expertise, these markets have achieved more substantial impacts than isolated efforts, showcasing the benefits of international cooperation.

Looking to the future, carbon trading markets are poised to evolve with technological and policy advancements. **Blockchain technology** offers promising avenues for enhancing transparency and accountability in carbon markets. By providing a secure and immutable record of transactions, blockchain can prevent fraud and ensure the integrity of emissions data. This technology could facilitate the expansion of carbon markets to include more sectors and countries, creating a truly global system. As more regions adopt carbon trading, the potential for linking markets increases, allowing for greater efficiency and coordination. Furthermore, policy innovations can address existing challenges, such as setting more ambitious caps and enhancing verification processes for offsets. By leveraging these advancements, carbon markets can play a critical role in driving global emissions reductions and fostering sustainable development.

9.3 Learning from the Global South: Adaptation and Mitigation

In the vast tapestry of climate impacts, the **Global South** stands out with its unique vulnerabilities. These regions, encompassing many developing countries, face an array of climate challenges that are distinct from those experienced elsewhere. Climate-induced migration is a pressing issue, as rising temperatures, sea levels, and extreme weather events force communities to leave their homes in search of safer grounds. This migration brings about social and economic strains, both for the regions they leave and those they move to. Moreover, agriculture, the backbone of economies in these areas, suffers immensely from climate disruptions. Unpredictable weather patterns and prolonged droughts threaten food security while dwindling water resources exacerbate the struggle for survival. These challenges underscore the urgent need for effective adaptation and mitigation strategies tailored to the local context.

In response to these challenges, the Global South has developed innovative adaptation strategies that draw from both traditional wisdom and modern science. Community-based adaptation projects have become a cornerstone of resilience efforts. These initiatives empower local communities to design and implement solutions that address their specific needs and conditions. By involving community members in the planning process, these projects ensure that adaptation measures are culturally appropriate and sustainable. Indigenous knowledge plays a vital role in these efforts, offering insights into resource management that have been honed over generations. For instance, traditional water harvesting techniques and rotational farming practices have proven effective in managing scarce resources and maintaining ecological balance. By integrating indigenous practices with scientific research, these projects enhance resilience against climate impacts.

Mitigation efforts in the Global South also showcase remarkable progress, particularly in the realm of renewable energy adoption. Solar energy projects across Africa exemplify the potential for sustainable energy solutions to drive emissions reductions and support economic development. Initiatives like the African Solar Industry Association have facilitated the deployment of solar panels in rural areas, bringing clean electricity to communities that previously relied on fossil fuels. These projects not only reduce emissions but also create jobs and improve livelihoods by providing reliable energy sources. In Southeast Asia, reforestation initiatives demonstrate another successful approach to mitigation. By restoring degraded forests and planting new trees, these efforts sequester carbon dioxide and enhance biodiversity. Reforestation projects, often spearheaded by local NGOs and supported by international partners, highlight the power of nature-based solutions in addressing climate change.

The importance of knowledge exchange and support cannot be overstated when considering the success of these initiatives. South-South cooperation frameworks enable countries in the Global South to share experiences, technologies, and best practices. By collaborating with one another, these nations can leverage their collective expertise to address shared challenges more effectively. International funding for adaptation projects further bolsters these efforts, providing the financial resources needed to implement large-scale initiatives. Organizations like the Adaptation Innovation Marketplace (AIM) connect local solutions with global financial communities, facilitating access to technical expertise and strategic partnerships. Through these collaborative networks, the Global South not only strengthens its resilience but also contributes valuable insights to the global climate discourse.

As the world continues to grapple with the impacts of climate change, the experiences and innovations of the Global South offer important lessons for all regions. By developing locally led solutions that address specific vulnerabilities, these countries demonstrate the potential for effective adaptation and mitigation strategies. The integration of traditional knowledge with modern science, coupled with international cooperation and support, creates a robust framework for building resilience and reducing emissions. These efforts not only benefit the communities directly involved but also contribute to global climate goals, highlighting the interconnectedness of our shared future.

9.4 Multilateral Organizations: Roles and Responsibilities

Multilateral organizations are the backbone of international climate governance, acting as key facilitators of global cooperation and policy implementation. The **United Nations Framework Convention on Climate Change (UNFCCC)** is at the forefront of these efforts, providing a platform for nations to come together and negotiate comprehensive climate agreements. Through its annual conferences, the UNFCCC fosters dialogue and consensus-building among member states, ensuring that all voices are heard in the fight against climate change. The Intergovernmental Panel on Climate Change (IPCC) complements this work by offering authoritative scientific assessments that guide policy decisions. The IPCC's reports synthesize the latest research, providing a clear picture of climate dynamics and the urgent need for action. Together, these organizations play a crucial role in shaping the global climate agenda, driving collective efforts to address the pressing challenges of our time.

The effectiveness of multilateral efforts in driving climate action is a mixed picture. On the one hand, these organizations have successfully facilitated the implementation of global climate agreements, such as the Paris Agreement, which brought nations together in a historic commitment to limit global warming. Through the establishment of mechanisms like the **Green Climate Fund**, multilateral bodies have mobilized financial resources to support climate initiatives, particularly in developing countries. However, achieving consensus among the diverse array of member states remains a significant challenge. Differing national interests and priorities can hinder progress, leading to lengthy negotiations and compromises that may dilute the effectiveness of agreements.

Despite these obstacles, the continued engagement of multilateral organizations is vital, as they provide the structure and legitimacy needed to coordinate international climate efforts.

Key initiatives and programs spearheaded by multilateral organizations demonstrate their capacity to drive meaningful change. The Green Climate Fund, for example, has launched numerous projects aimed at enhancing climate resilience and reducing emissions in vulnerable regions. These initiatives focus on areas such as renewable energy deployment, sustainable agriculture, and ecosystem restoration, offering tangible solutions to climate challenges. Similarly, the United Nations Environment Programme (UNEP) leads climate adaptation programs that prioritize capacity-building and knowledge transfer. These programs equip communities with the tools and skills needed to adapt to changing environmental conditions, fostering resilience and sustainability. By leveraging their global reach and expertise, multilateral organizations are able to implement impactful projects that address both the causes and effects of climate change.

Looking to the future, multilateral organizations must evolve to meet emerging challenges and enhance cooperation. Expanding their focus on climate justice and equity will be essential in ensuring that the benefits of climate action are shared fairly among all countries and communities. By integrating considerations of social and economic justice into their frameworks, these organizations can address the disproportionate impacts of climate change on marginalized populations. Additionally, the integration of new scientific findings into policy frameworks will be crucial in keeping pace with the rapidly evolving landscape of climate science. As research continues to uncover new insights into climate dynamics, multilateral bodies must adapt their strategies to reflect the latest knowledge. This flexibility will enable them to respond effectively to emerging threats and opportunities, ensuring that international climate efforts remain relevant and impactful.

In navigating the complexities of global climate governance, multilateral organizations stand as pillars of cooperation and progress. Their ability to unite nations under a common cause while addressing diverse needs and priorities is a testament to their enduring significance. As we look ahead to the future, the role of these organizations will remain pivotal in guiding the world toward a sustainable and equitable future, where climate action is both a shared responsibility and a collective achievement.

Reflection Section

Consider how global cooperation impacts climate action. Reflect on the role of international agreements in addressing climate challenges. How do these efforts resonate with your personal views on climate change? Explore this idea with a journaling prompt: Describe how international collaboration influences your perspective on climate action. How does it shape your understanding of the challenges and opportunities in creating a sustainable future? How can countries agree on a trillion-dollar expenditure to reverse global warming unless a clear picture, a proof-of-principle, and a guarantee that it will not fail?

Chapter 10

Addressing Common Objections

Imagine a world where every small action you take can ripple outwards, creating waves of change across the globe. This is the transformative power of individual action in the fight against climate change. It's easy to feel overwhelmed by the magnitude of the climate crisis and doubt the impact of one person's efforts. Many believe that individual actions are too small to matter, but this mindset underestimates the collective influence we hold. When millions of people make small changes, the cumulative effect can drive significant environmental benefits. Consider the simple act of choosing a reusable water bottle over a disposable one. Alone, it may seem insignificant, but multiplied by millions, it reduces plastic waste and decreases the demand for single-use plastics, prompting companies to innovate and offer more sustainable options.

11.1 Consumer Choice

Consumer choice wields tremendous power in driving market change. As individuals increasingly opt for eco-friendly products, businesses respond by developing greener alternatives. This shift is not just a reaction to consumer demand but a recognition of the growing market for sustainable goods. When people choose products that align with their environmental values, they signal to companies that sustainability is a priority, encouraging them to adopt more responsible practices. This dynamic is evident in the rise of organic foods, electric vehicles, and energy-efficient appliances. As these products become mainstream, they drive competition and innovation, leading to more accessible and affordable options for everyone.

Grassroots movements further highlight the power of collective action. Across the world, communities have united to advocate for policies that address climate change. These movements, often initiated by passionate individuals, have successfully influenced local and national legislation. For instance, local recycling programs have emerged as a result of citizen-led initiatives, demonstrating the impact of community-driven efforts. These programs not only reduce waste but also raise awareness about the importance of recycling, inspiring others to participate. Similarly, energy conservation efforts in residential neighborhoods have led to significant reductions in energy use, proving that collective action can drive meaningful change. Seeing the National Academy Press (NAP.edu) display a radical decrease agenda from 50 Gt/yr emission rate down to zero by 2100 is quite remarkable and unrealistic no doubt, where GHG = Green House Gas emissions.

IEEE ISTAS 2019   Negative Emissions and Carbon Sequestration

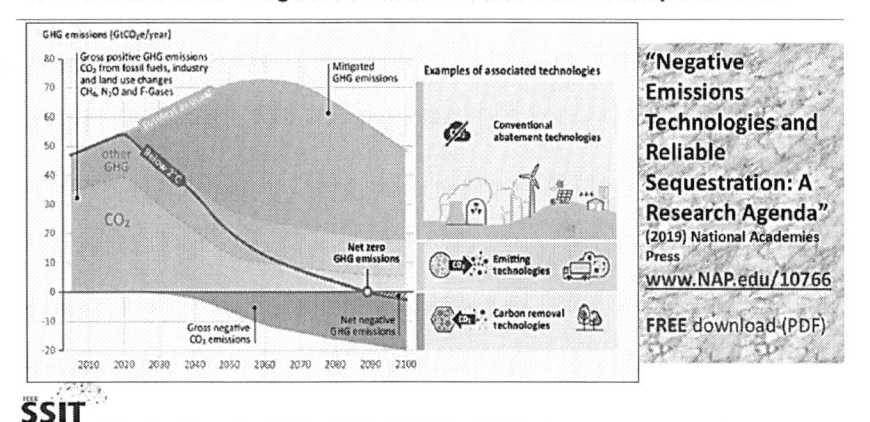

"Negative Emissions Technologies and Reliable Sequestration: A Research Agenda" (2019) National Academies Press www.NAP.edu/10766

FREE download (PDF)

The ripple effect of personal responsibility is a powerful force for change. One person's actions can inspire others, creating a chain reaction that amplifies their impact. This phenomenon is particularly evident on social media, where individuals share their sustainable habits and encourage others to follow suit. Social media campaigns highlighting simple changes, like reducing meat consumption or conserving energy, have gained traction, reaching a wide audience and sparking broader adoption of these practices. These campaigns not only raise awareness but also foster a sense of community among those committed to sustainability, reinforcing the idea that individual efforts matter.

Common objections to personal impact often stem from the belief that individual actions are too small to make a difference. However, statistical evidence shows that the cumulative effect of individual actions can be profound. For instance, if every household in the U.S. replaced just one incandescent bulb with an LED, it would save enough energy to power millions of homes for a year. This example illustrates how small, everyday actions, when multiplied across a population, can lead to significant environmental benefits. By refuting the notion that individual actions are negligible, we empower ourselves to take responsibility for our impact and inspire others to do the same.

The most radical objection to large-scale CDR was recently published with the subtitle, "What is Sustainable and Just?" However, neither author is a scientist but instead expresses exaggerated political ethics without any concern for the survival of the species. For example, they claim the use of CDR will be "...breaching planetary boundaries [or] human rights...", [with] "...serious consequences for climate risk, sustainability, justice, and geopolitics." This obviously ignores the much bigger climate risk of the coming hothouse earth with a lot of superficial objections. See McLaren, Duncan & Olaf Corry, "Carbon Dioxide Removal: What is Sustainable and Just?", Environment: Science and Policy for Sustainable Development, Vol. 67, Issue 1, Feb. 2025. Open access.

In this interconnected world, every choice we make sends a message. It tells companies what we value, it tells policymakers what we prioritize, and it tells others what we stand for. The power of individual action lies not just in its immediate impact but in its ability to inspire change at all levels. As we continue to address the climate crisis, let us remember that each of us has a role to play. Together, our individual actions create a collective force that can drive the systemic changes needed for a sustainable future.

11.2 Feasibility of Large-Scale Gigaton Carbon Technologies

In the realm of climate solutions, carbon capture and storage (CCS) technologies stand as a beacon of hope, promising to curb emissions on a massive scale. Today, operational CCS facilities, like Norway's Sleipner project, successfully sequester millions of tons of CO2 annually, paving the way for broader adoption. These facilities capture emissions at their source, such as power plants, and store them underground, preventing their release into the atmosphere. The potential for growth in this sector is immense, with advancements in technology and infrastructure promising more efficient and cost-effective solutions. Successful pilot projects, like **Climeworks'** Orca plant in Iceland, have demonstrated the viability of direct air capture, transitioning these innovative concepts from pilot to commercial use. These projects not only capture carbon directly from the air but also highlight the scalability of such technologies, proving that with the right investment and support, gigaton-level carbon removal is within reach, such as with Carbon Engineering DAC shown here.

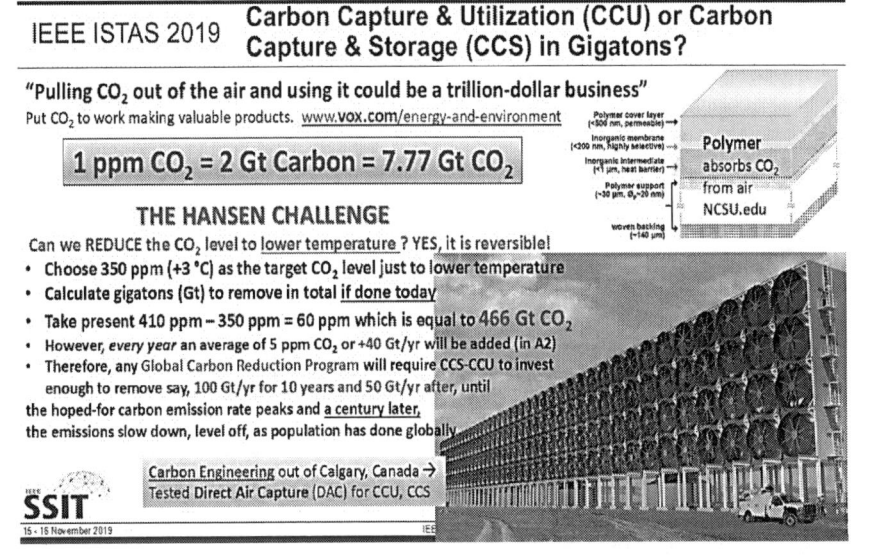

IEEE ISTAS 2019 — Carbon Capture & Utilization (CCU) or Carbon Capture & Storage (CCS) in Gigatons?

"Pulling CO$_2$ out of the air and using it could be a trillion-dollar business"
Put CO$_2$ to work making valuable products. www.vox.com/energy-and-environment

1 ppm CO$_2$ = 2 Gt Carbon = 7.77 Gt CO$_2$

Polymer cover layer (<500 nm, permeable)
Inorganic membrane (<200 nm, highly selective)
Inorganic intermediate (<1 μm, heat barrier)
Polymer support (~30 μm, 0$_2$~20 nm)
woven backing (~140 μm)

Polymer absorbs CO$_2$ from air
NCSU.edu

THE HANSEN CHALLENGE
Can we REDUCE the CO$_2$ level to lower temperature ? YES, it is reversible!
- Choose 350 ppm (+3 °C) as the target CO$_2$ level just to lower temperature
- Calculate gigatons (Gt) to remove in total if done today
- Take present 410 ppm – 350 ppm = 60 ppm which is equal to 466 Gt CO$_2$
- However, every year an average of 5 ppm CO$_2$ or +40 Gt/yr will be added (in A2)
- Therefore, any Global Carbon Reduction Program will require CCS-CCU to invest enough to remove say, 100 Gt/yr for 10 years and 50 Gt/yr after, until the hoped-for carbon emission rate peaks and a century later, the emissions slow down, level off, as population has done globally

Carbon Engineering out of Calgary, Canada →
Tested Direct Air Capture (DAC) for CCU, CCS

SSIT
15 - 16 November 2019

Yet, the road to scaling these technologies is fraught with economic and logistical challenges. The initial investment required for CCS and direct air capture is substantial, often acting as a barrier to widespread implementation. However, recent trends show a positive shift, with increased investment in clean technology sectors.

Selling carbon credits for carbon dioxide removal (CDR) has emerged as a promising avenue, providing financial incentives for companies to invest in carbon capture projects such as Climeworks. Innovations in technology are also playing a crucial role in reducing costs and improving efficiency. The goal is to achieve a cost of **less than $10 per ton** of CO_2 removed ($10 billion per gigaton CDR) making these solutions more accessible and appealing to industry stakeholders. With continued support and investment, these hurdles can be overcome, driving the widespread adoption of CCS technologies.

IEEE ISTAS 2019 **Transformative Zero or Negative Emissions Tech**

"A Process for Capturing CO_2 from the Atmosphere"
DAVID KEITH ET AL., JOULE, VOLUME 2, ISSUE 8, P1573-1594, AUGUST 15, 2018

→ Estimates low cost can be around $100/ton of CO_2 presently
→ $50 billion/yr for 50 Gt/yr @$1/t

Carbon Engineering – very low-carbon fuels, powered by renewables, using CO_2 from the air, drawing hydrogen from electrolysis to produce hydrocarbons. The company calls the process "air to fuels," or A2F, and it is targeting wide commercialization in 2021.

HyTech is targeting a big market – diesel engines – the source of 50% of urban smog, especially in winter

Onboard electrolyzers are the game plan for turning existing → diesel engine fleet into **zero-emissions vehicles (ZEV)** by making them run on pure hydrogen.
-- HyTech Power, based in Redmond, Washington

Recent technological advancements have further enhanced the feasibility of carbon capture and storage. Breakthroughs in material science, such as the use of olivine for large-scale CDR, have shown promising results in increasing carbon capture efficiency. Organizations like **Carbon180.org** are at the forefront of these innovations, developing cutting-edge solutions that push the boundaries of what's possible. Highlighting the best carbon capture technologies, those with the most rapid scale-up goals and low ongoing costs provides a roadmap for the future. Estimates for capital investment in solar and wind-powered gigaton CDR projects suggest a sustainable path forward, combining renewable energy with carbon capture to create a holistic approach to emissions reduction. The integration of renewable power not only reduces the carbon footprint of these operations but also ensures their sustainability in the long term.

Common misconceptions about the limitations of large-scale carbon technologies often stem from a lack of understanding of their potential. Critics argue that such technologies are impractical, citing high costs and complex logistics. However, history has shown us that seemingly insurmountable challenges can be overcome. Consider the transatlantic cable, which revolutionized global communication, or the Apollo moon missions, which pushed the boundaries of human exploration. The development of electric vehicles and hydrogen power faced similar skepticism, yet today they are integral components of our sustainable future. The effectiveness of contests, such as the XPRIZE, in fostering technological breakthroughs highlights the power of innovation. Moreover, the involvement of billionaires, like those behind SpaceX and Blue Origin, demonstrates the potential of private investment to drive progress. Their commitment to advancing carbon capture technologies underscores the feasibility of achieving gigaton-scale carbon removal, proving that with vision and determination, the future of sustainable energy is within our grasp.

11.3 Overcoming Eco-Anxiety: Survival-Focused Narratives

Eco-anxiety is a term that has gained traction as climate change becomes an ever-pressing issue. This form of anxiety stems from the fear and uncertainty surrounding global warming and its potential consequences. People, especially the younger generation, often feel a profound sense of hopelessness about the future. They worry about rising sea levels, extreme weather events, and the long-term viability of life on Earth. This anxiety can manifest in various ways, from chronic worry and despair to a general feeling of helplessness. It's a psychological burden that weighs heavily on individuals and communities, affecting mental well-being and everyday life. The constant news of environmental degradation and the seemingly slow pace of systemic change can exacerbate these feelings, leading to a pervasive sense of doom.

However, there are practical strategies to manage eco-anxiety and foster a positive outlook. Mindfulness and stress-reduction techniques, such as meditation and deep breathing exercises, can help individuals stay grounded and focused on the present. These practices encourage a shift away from catastrophic thinking and towards a more balanced perspective. Engaging in proactive environmental activities, like participating in local clean-up events or supporting renewable energy projects, provides a sense of agency and purpose, counteracting feelings of helplessness. By actively contributing to solutions, individuals can transform their anxiety into motivation. Additionally, studying the most promising gigaton capture technologies offers a glimpse of hope. Understanding the potential of these solutions to reverse global temperature trends can reassure people that effective actions are underway.

Storytelling plays a crucial role in creating hope and alleviating eco-anxiety. Narratives that focus on solutions and highlight success stories can shift the narrative from one of despair to one of possibility. Communities worldwide are demonstrating resilience and innovation in the face of climate challenges. For example, cities implementing clean energy solutions or local groups restoring ecosystems provide tangible examples of positive change. Media that emphasizes these successes rather than solely focusing on problems can help reshape the public discourse around climate change. By sharing stories of technological advancements and community-led initiatives, we can inspire others to believe in the potential for a sustainable future.

Focusing on actionable solutions rather than dwelling on problems is essential for empowerment. Community support groups and climate action forums offer spaces for individuals to connect, share experiences, and collaborate on projects. These platforms foster a sense of solidarity and collective action, reinforcing the idea that no one is alone in their concerns.

Educational programs emphasizing proactive solutions, such as white roof initiatives, reflective metal roofs, and phase change insulation, demonstrate practical ways to address climate issues. These programs provide individuals with the knowledge and tools needed to make informed decisions and take meaningful action. By channeling energy into constructive efforts, we can move from fear to empowerment, creating a mindset focused on hope and possibility.

11.4 Balancing Daily Life with Climate Action

Incorporating sustainable practices into your daily routine doesn't require an overhaul of your lifestyle. Using less utility-supplied, natural gas or propane-driven electricity can be a beginning. We recommend solar energy lights, such as MpowerD.com products in those dark places around the house and office.

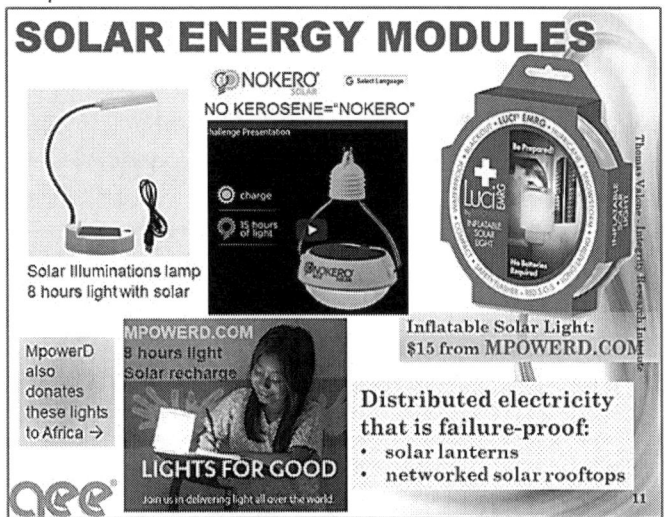

SOLAR ENERGY MODULES

NOKERO
NO KEROSENE="NOKERO"

Solar Illuminations lamp
8 hours light with solar

MpowerD also donates these lights to Africa →

MPOWERD.COM
8 hours light
Solar recharge
LIGHTS FOR GOOD
Join us in delivering light all over the world

Inflatable Solar Light:
$15 from MPOWERD.COM

Distributed electricity that is failure-proof:
- solar lanterns
- networked solar rooftops

You may also want to add meal planning using sustainable ingredients. This means opting for locally sourced produce and reducing meat consumption. Supporting vertical farming companies can also make a difference; they use less land and water, offering a more sustainable food source. Additionally, lab-grown meat should be considered as an alternative. These practices not only reduce your carbon footprint but also promote biodiversity and healthier ecosystems. Integrating these small changes into your meal planning can have a big impact over time, both environmentally and on your health.

Energy consumption is another area where small changes can add up. Utilizing smart home technology, like programmable thermostats and energy-efficient appliances, can significantly reduce your energy use. Installing solar panels or wind turbines, even on a small scale, can further decrease your reliance on fossil fuels. These technologies are becoming more accessible and affordable, and they offer long-term savings on energy bills. Adding solar-powered lights to closets (e.g., Luci Solar Outdoor Lantern by **MpowerD.com**) and other small spaces is another simple step towards sustainability. These incremental changes not only reduce your carbon footprint but also create a more energy-efficient home. We use them all over the house.

The transition to **a plant-based diet** is also one of the most impactful changes you can make. It uses substantially less land and water compared to a meat-animal based diet, reducing environmental degradation. Start by incorporating more plant-based meals into your week and gradually increase them as you find recipes you enjoy. Not only is a meat-based diet excessively supporting vast land use for a limited number of animals for years but farming itself is land intensive.

Today, the stock prices of **VERTICAL farms** have skyrocketed. They use 90% less soil and water, with NO insecticides! Whole Foods Market carries vertical farming food products for sale. Similarly, using public transport or carpooling can significantly cut down on emissions. It's a practical choice that also reduces traffic congestion and air pollution. These small, incremental changes don't require drastic lifestyle shifts but can collectively lead to substantial environmental benefits. A summary of **AeroFarms** "Breakthrough Achievements" is listed below:

- World's largest
- 2 million pounds/year leafy green vegetables
- No soil, No pesticides
- 390x productivity
- **95% less water**

Despite the benefits, sustainable living often faces obstacles. Budget constraints are a common barrier, as eco-friendly products can be more expensive. To overcome this, look for budget-friendly options like bulk buying or choosing store brands. Time management is another challenge, as busy schedules can make it difficult to prioritize sustainable practices. Planning meals, combining errands to reduce travel, and setting aside time for eco-friendly activities can help. By addressing these barriers, you make sustainable living more feasible and less daunting.

The long-term benefits of adopting sustainable practices are significant. Reducing energy consumption and opting for cleaner transportation methods contribute to lower air pollution, leading to better health outcomes. Cleaner air reduces the incidence of respiratory illnesses and improves overall well-being. Financially, sustainable practices often result in savings. Energy efficiency upgrades, like better insulation and energy-efficient appliances, lower utility bills over time. These savings can be reinvested into further sustainability efforts, creating a positive feedback loop. By adopting these practices, you not only contribute to a healthier planet but also improve your quality of life and future-proof your finances.

In weaving these changes into your daily life, you align personal habits with the broader goal of sustainability. It's about finding balance and making conscious choices that reflect your values. Each step you take contributes to a larger movement towards a sustainable future, demonstrating that meaningful change is possible through collective individual efforts. While the journey to sustainable living is ongoing, every change you make brings us closer to a healthier, more resilient world. As we explore strategies in the next chapter further, consider how these practices can be integrated into your own life, enhancing both personal and environmental well-being.

Reflection Section

Take a moment to consider the small changes you've made in your life to reduce your carbon footprint. Write down three actions you've taken and reflect on how they contribute to the larger goal of sustainability. Consider sharing your actions with friends or on social media to inspire others to join the movement.

Chapter 11

Education and Advocacy for Systemic Change

Imagine standing at a crossroads, where one path leads to the status quo and the other to a future shaped by informed action and systemic change. Education and advocacy are the compasses guiding us toward the latter, equipping us with the tools to influence policy and public perception. In a world saturated with information, effective climate communication is pivotal. It's not just about data but how we present that data, transforming it into compelling narratives that resonate with diverse audiences. Communication is an art in itself, a dance between facts and feelings, where every word and image can tip the scale toward awareness and action.

11.1 Effective Climate Communication Strategies

At the heart of impactful climate advocacy lies clear, relatable, and persuasive communication. The principles of effective climate communication revolve around simplifying complex ideas without diluting their essence. Imagine explaining the intricacies of climate change to a friend over coffee. You'd avoid jargon and instead use simple language to convey your point. This is crucial in making scientific concepts digestible. Tailoring messages to resonate with different audiences is equally important. For a high school student, the message might focus on the future they will inherit, while for local policymakers, it might highlight economic benefits. Each audience requires a unique approach, ensuring the message hits home.

Science communication plays a vital role in advocacy by translating dense data into accessible information. Visual aids like infographics can distill complex statistics into engaging visuals, making them easier to understand and retain. Imagine a chart that compares CO_2 emissions over the decades, instantly conveying the urgency of the situation. Storytelling techniques can humanize scientific findings, turning them into relatable narratives. Picture a documentary that follows a farmer adapting to climate changes, grounding abstract facts in real-world experiences. These tools bridge the gap between knowledge and action, empowering audiences to grasp the implications of climate data and inspiring them to support climate campaigns actively.

Framing issues and crafting compelling narratives can significantly influence public perception and policy. When we frame climate action as an opportunity rather than a burden, we shift the conversation from sacrifice to potential. It's not just about cutting emissions; it's about building resilient communities and fostering innovation. This approach reframes challenges as catalysts for growth, engaging audiences who might otherwise feel overwhelmed by the magnitude of climate change. By crafting narratives that emphasize positive outcomes, like cleaner air and healthier communities, we can inspire collective action and drive policy changes that prioritize sustainability.

Engaging diverse audiences requires strategies that foster inclusive dialogue. Culturally relevant communication approaches ensure that messages resonate with varied demographics. This might involve using local languages or incorporating cultural symbols that hold significance for the audience. Social media platforms expand our reach, allowing us to connect with global audiences and share climate messages widely. Consider a viral campaign that uses hashtags to promote sustainable practices, engaging users from different backgrounds in a shared conversation. By adapting our communication strategies to meet the needs of different audiences, we create a more inclusive dialogue that encourages diverse voices to join the climate conversation.

11.2 Mobilizing for Policy Change: Tools and Tactics

Navigating the labyrinth of policy change is a task that demands both strategy and resources. At the forefront of this endeavor are tools that galvanize support for climate policy shifts. Online petition platforms, like Change.org, serve as a digital megaphone, amplifying the voices of individuals and transforming them into a collective roar that policymakers cannot ignore. These platforms allow you to create petitions that address specific climate issues, gathering signatures from thousands, sometimes millions, of supporters worldwide. The impact of these petitions lies in their ability to demonstrate public demand for change, providing a tangible metric that can sway decision-makers.

In addition to online platforms, advocacy toolkits and guides are invaluable resources. These comprehensive documents equip advocates with the necessary knowledge and skills to communicate their goals effectively. Toolkits often include templates for letters, talking points, and strategies for engaging with media outlets. They serve as a roadmap for organizing campaigns and offer practical advice on navigating political landscapes. With these resources, you can craft compelling messages and coordinate actions that resonate with both the public and policymakers. By leveraging these tools, you transform passion into actionable steps, bridging the gap between grassroots energy and legislative change.

Successful tactics for influencing policy are as varied as they are effective. Organizing letter-writing campaigns can be a powerful method for directly reaching policymakers. By mobilizing community members to send personalized letters to their representatives, you create a wave of communication that highlights the urgency and importance of climate issues. Each letter serves as a personal testament, urging lawmakers to prioritize environmental concerns. Additionally, coordinating public demonstrations and rallies can capture media attention and generate public pressure. These events, whether they involve marching through city streets or gathering in public squares, create a visual representation of community commitment to climate action, capturing the attention of those in power.

Grassroots mobilization plays a crucial role in driving policy shifts both nationally and internationally. Local movements, driven by passionate individuals and communities, can generate significant pressure on governments to enact change. Community organizing and canvassing are foundational elements of grassroots efforts. By engaging with neighbors and organizing local events, you can build a network of advocates who share your vision for a sustainable future. These localized efforts can snowball, influencing broader policy discussions and inspiring similar movements in other regions. Grassroots mobilization fosters a sense of collective responsibility and empowers individuals to become catalysts for change.

A roadmap for engaging with policymakers is essential for effective advocacy. Building relationships with decision-makers requires strategic communication and persistence. Scheduling meetings with local representatives allows you to present your case directly, fostering a dialogue that can lead to meaningful policy changes. During these meetings, it's important to articulate your concerns clearly and concisely, providing evidence and solutions that resonate with the representative's values and priorities. Preparing policy briefs and position papers can support your arguments, offering detailed analysis and recommendations. These documents serve as a reference for policymakers, ensuring that your message remains front and center in their decision-making process.

When approaching policymakers, remember that they are human too. Building rapport and establishing trust can be as important as the policy details themselves. Approach them with respect and openness, be ready to listen, and advocate. Anticipate their questions and prepare to address any concerns they might have. By framing your message in terms of shared goals and benefits, you can create a collaborative atmosphere where both parties work toward common objectives. This approach not only increases the likelihood of policy change but also fosters long-term relationships that can support future advocacy efforts.

In the realm of policy change, every voice counts. Your efforts, whether through an online petition or a face-to-face meeting, contribute to a larger movement pushing for a sustainable and equitable future. By utilizing the tools and tactics available, you amplify your Impact, ensuring that climate policy remains a priority on the legislative agenda. As you engage in this work, remember that you are part of a global community of advocates, each playing a role in shaping the policies that will define our planet's future.

11.3 Building Climate Alliances: Networking for Impact

In the realm of climate advocacy, building alliances is not just beneficial; it's transformative. When individuals and organizations unite, their collective power can amplify efforts, achieving impacts that would be unimaginable alone. Collaborating with environmental NGOs like EESI.org, the Rocky Mountain Institute, or carbon180 allows advocates to pool resources, share expertise, and coordinate strategies. These organizations bring specialized knowledge and networks that can enhance the effectiveness of climate initiatives. By forming coalitions with like-minded groups, you can create a united front that draws on diverse strengths, making your advocacy efforts more robust and far-reaching. This collective approach ensures that the message is consistent and the Impact is multiplied.

Networking effectively within the climate movement requires intentionality. Attending climate conferences and workshops is a strategic way to establish connections with other advocates, researchers, and policymakers. These events provide opportunities to learn from experts, exchange ideas, and collaborate on projects. Imagine the insights gained from a conversation with a researcher who shares new data on carbon sequestration or a policy expert who reveals effective lobbying techniques. Similarly, joining professional networks and online forums dedicated to climate issues allows for continuous engagement and knowledge sharing. These platforms are invaluable for staying informed about the latest developments and best practices. By participating in these networks, you build relationships that can support and strengthen your advocacy efforts over time.

The benefits of cross-sector collaboration are immense. Partnerships with businesses can drive sustainable practices by integrating environmental considerations into corporate strategies. Companies often have the resources and influence to implement large-scale solutions, and when they collaborate with climate advocates, the potential for meaningful change increases exponentially. Engaging academic institutions for research support can also be incredibly beneficial. Universities and research centers offer cutting-edge research capabilities and analytical tools that can provide the evidence needed to drive policy changes. By working together, these sectors can create innovative solutions to complex climate challenges, mobilizing resources that would otherwise remain untapped.

Consider the successful climate alliances that have driven significant policy changes or initiatives. The Global Covenant of Mayors for Climate & Energy is a prime example of how cities around the world can unite to tackle climate change. This coalition brings together local governments committed to reducing greenhouse gas emissions and enhancing resilience to climate impacts. By sharing strategies and data, member cities can learn from one another and implement successful programs in their own jurisdictions. Another noteworthy alliance is the Climate Action Network, which coordinates campaigns across continents, bringing together over 1,300 NGOs to advocate for sustainable policies. Their collaborative efforts have influenced international climate negotiations and pushed for stronger commitments under agreements like the Paris Accord.

11.4 Visual Element: Networking Map

Imagine a visual map illustrating the connections between various climate organizations, coalitions, and sectors. This map serves as a guide to understanding how these entities interact and collaborate, showcasing the web of relationships that drive climate advocacy.

Such a tool can inspire you to identify potential partners and strategize your networking efforts within this interconnected ecosystem.

Building alliances in climate advocacy is about more than just making connections; it's about creating a community dedicated to a common cause. By understanding the importance of these relationships and actively engaging in networking activities, you can become an integral part of a movement that drives systemic change. Each alliance strengthens the collective voice of climate advocates, ensuring that their efforts resonate on a global scale. Whether through collaboration with NGOs, partnerships with businesses, or support from academia, these alliances provide the foundation for innovative solutions and impactful policy changes. By participating in this collaborative effort, you contribute to a legacy of resilience and sustainability that will shape the future.

11.5 The Power of Storytelling in Climate Advocacy

Storytelling is a potent tool in shaping public opinion and policy, especially when it comes to climate advocacy. Personal and compelling narratives have a unique ability to engage emotions, creating a connection that facts and figures alone often cannot achieve. When individuals share their personal stories of how climate change has impacted their lives, it paints a vivid picture that statistics can rarely match. These stories bring to life the abstract concepts of rising sea levels and increasing temperatures, making them tangible and relatable. Consider the tale of a coastal community grappling with the encroachment of the sea or a farmer facing dwindling yields due to changing weather patterns. These narratives resonate deeply, prompting those who hear them to reflect on the broader implications of climate change and the need for action.

Crafting impactful climate stories requires a thoughtful approach. A well-structured narrative should have a clear beginning, middle, and end, guiding the audience through a journey that is both informative and engaging. The beginning sets the scene, introducing the characters and the challenges they face. The middle delves into the struggle, highlighting the obstacles and conflicts encountered. The end offers resolution, whether it's a call to action or an inspiring solution. Incorporating emotional appeal and human interest elements can further enhance the story's Impact. By focusing on individual experiences and emotions, these narratives become relatable and memorable, encouraging audiences to empathize with those affected by climate change and to consider their own roles in addressing it.

The integration of multimedia in storytelling greatly enhances the delivery and reach of climate advocacy narratives. Short films and documentaries offer a powerful medium to convey complex stories with immediacy and Impact. Visuals, combined with sound and narrative, create a multisensory experience that captivates audiences. Campaigns like "Years of Living Dangerously" and "Six Degrees Could Change the World" have successfully used this format to educate viewers on the realities of climate change, influencing both public perception and policy discussions. These productions not only inform but also inspire, driving viewers to seek out more information and engage in climate action.

Utilizing social media for storytelling campaigns can expand the reach of climate narratives, engaging a global audience. Platforms like Instagram, Twitter, and TikTok offer creative avenues for sharing stories through videos, images, and interactive content. A well-crafted social media campaign can rapidly gain traction, spreading awareness and encouraging engagement. By leveraging the power of multimedia and social networks, climate advocates can reach diverse demographics, from teenagers to policymakers, fostering a broader conversation about climate change and its impacts.

Successful uses of storytelling in climate advocacy abound, providing valuable lessons for those seeking to influence change. The narrative-driven approach of the documentary series "An Inconvenient Truth" played a pivotal role in raising awareness about global warming, ultimately contributing to shifts in both public opinion and policy. Similarly, interactive storytelling platforms like the online game "World Climate Simulation" allow users to explore the complexities of climate negotiations, offering an engaging and educational experience that emphasizes the importance of international cooperation. These examples demonstrate how storytelling can be a catalyst for change, translating the abstract into the tangible and motivating individuals to take action.

As we explore the power of storytelling, it becomes clear that narratives have the capacity to transcend barriers, connecting with audiences on a personal level. By employing storytelling techniques, climate advocates can amplify their messages, engaging hearts and minds in the fight against climate change. Whether through personal stories, multimedia campaigns, or interactive platforms, storytelling is a vital component of effective climate advocacy, fostering awareness, empathy, and action.

Reflection Section

Craft your climate message by taking a moment to consider the audience you wish to engage in climate advocacy. Reflect on their values, interests, and concerns. Now, craft a brief message that speaks directly to them, using simple language and relatable examples. Consider how you incorporate visual elements to enhance your message. This exercise helps you practice tailoring your communication to resonate with diverse audiences, a crucial skill in effective climate advocacy.

Chapter 12

Evaluating and Improving Carbon Removal Strategies

Imagine for a moment a vast network of invisible threads, each representing a unit of carbon dioxide captured from the atmosphere and stored safely away. These threads, though unseen, are critical to weaving a future where our planet thrives. In this chapter, we explore the intricate process of evaluating and improving carbon removal strategies, which is essential for mitigating climate change. The key to success lies in the metrics and measurements that guide these efforts, ensuring that our actions are not merely hopeful but scientifically grounded and effective.

12.1 Metrics and Measurements: Assessing Impact

In the realm of carbon removal, standardized metrics are the backbone of progress. They provide the consistency needed to evaluate the effectiveness of various projects, ensuring that efforts are both accountable and impactful. Without these standards, comparing the success of different initiatives would be like trying to measure the length of shadows without a ruler. The development of carbon accounting standards plays a crucial role here. These standards consider both capital investment and ongoing maintenance costs, helping us understand the true economic footprint of carbon removal technologies. Lifecycle analysis further enriches this understanding, offering insights into the entire lifespan of a project—from the initial resource extraction to the final stages of decommissioning. This holistic view is vital for identifying areas where efficiency can be improved or costs reduced, making carbon removal not only feasible but sustainable in the long run.

The metrics used to assess carbon removal impact are as varied as the projects themselves, yet they share common threads of measurement that ensure clarity and comparability. One of the most straightforward indicators is the number of tons of CO2 sequestered per year. This metric provides a concrete measure of a project's impact, allowing stakeholders to gauge its effectiveness in reducing atmospheric carbon. Another critical metric is the cost per ton of carbon removed, which highlights the economic efficiency of a project. This figure is essential for decision-makers who must balance environmental goals with budget constraints. By understanding these metrics, you gain insight into the nuanced dynamics of carbon removal, appreciating both the challenges and the triumphs that come with it.

Research Open

Geology, Earth and Marine Sciences
Volume 4 Issue 3

Research Article

Direct Air Capture and Removal of Gigatons of CO_2 Offers Hope for Climate Recovery

Thomas F. Valone PhD*

Integrity Research Institute

*Corresponding author: Thomas F. Valone, Integrity Research Institute
Received: October 21, 2022; Accepted: October 25, 2022

Asian Journal of Environment & Ecology
Volume 20, Issue 2, Page 42-58, 2023; Article no.AJEE.97404
ISSN: 2456-690X

Gigatonne Carbon Dioxide Removal Can Reverse Global Heating Trend

Thomas F. Valone [a*]

[a] Integrity Research Institute, 5020 Sunnyside Ave., Suite. 209, Beltsville MD 20705, US.

Author's contribution

The sole author designed, analysed, interpreted and prepared the manuscript.

Article Information

DOI: 10.9734/AJEE/2023/v20i2436

Advanced tools and technologies enhance the precision of these measurements, acting as the eyes and ears of the carbon removal industry. Satellite monitoring is a game-changer for large-scale projects, offering a bird's-eye view of carbon capture efforts across vast expanses. These satellites provide real-time data on atmospheric CO2 levels, helping to verify the effectiveness of sequestration activities. Sensor networks further refine this data collection, offering real-time insights into localized carbon dynamics. These networks, often deployed in forests or near industrial sites, provide continuous data that can be used to fine-tune carbon removal strategies. For a more personal touch, carbon footprint calculators offer project-level evaluations, allowing individual initiatives to assess their Impact and make informed adjustments.

Case studies of effective measurement practices offer valuable lessons and inspiration for future projects. The **Carbon Trust's** methodology for project evaluation sets a high standard, combining rigorous data analysis with practical insights to assess carbon removal initiatives. This approach ensures that projects are not only effective but also efficient, maximizing their Impact while minimizing resource use. Consider a forestry carbon offset project that employs a robust measurement system. By using satellite imagery and ground-based sensors, this project can accurately track the growth of trees and the amount of CO2 they sequester. This data is then used to calculate the project's overall carbon offset, providing transparent and verifiable results that can be shared with stakeholders. Such examples demonstrate the power of precise measurement in driving meaningful progress in carbon removal.

12.2 Continuous Improvement: Iterating on Carbon Strategies

The concept of continuous improvement is a cornerstone in the realm of carbon removal, much like the iterative processes that drive technological advancements across industries. At its core, continuous improvement involves making incremental adjustments to enhance the effectiveness and efficiency of carbon capture strategies. Think of it as fine-tuning a musical instrument, where each adjustment brings you closer to the perfect harmony of environmental balance. Feedback loops play a pivotal role here, acting as the mechanism that channels insights from each iteration back into the system. These loops enable project managers to assess the performance of current strategies and identify areas for refinement. By continuously feeding data back into the process, projects can evolve dynamically, adapting to new challenges and opportunities with agility and precision.

Refining carbon removal techniques requires a blend of innovation and structured methodology, borrowing best practices from fields like manufacturing and software development. **Lean Six Sigma**, a methodology known for its focus on reducing waste and improving quality, offers valuable insights for carbon projects. By applying Lean Six Sigma principles, teams can identify inefficiencies in carbon capture processes and implement targeted improvements. This approach not only enhances effectiveness but also maximizes resource utilization, ensuring that every effort contributes to meaningful carbon reduction. Similarly, agile methodologies provide a framework for responding to the ever-changing landscape of carbon removal. Agile practices encourage experimentation and learning, enabling teams to test new ideas and pivot quickly when needed. This flexibility is crucial in a field with constant technological advancements and high stakes.

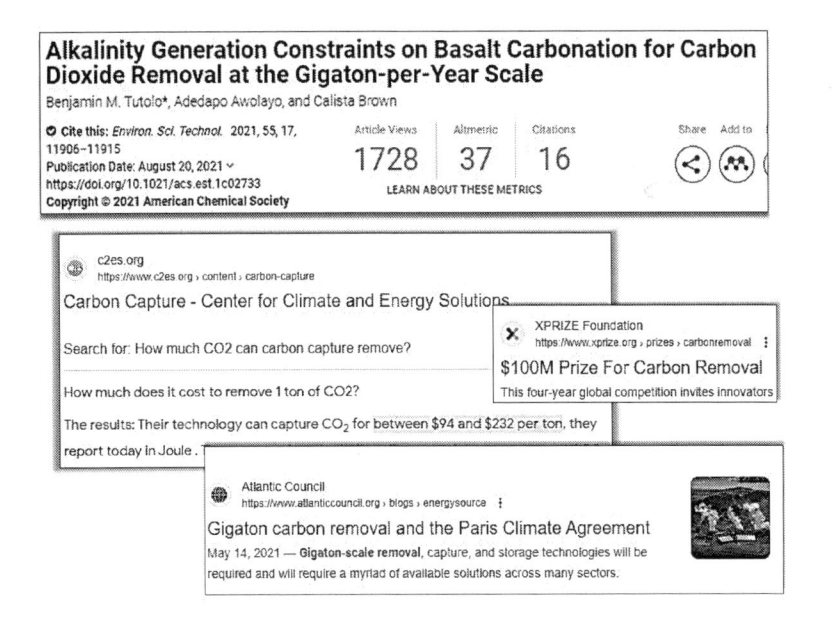

Data analysis serves as the backbone of strategy iteration, offering a wealth of insights that drive informed decision-making. The advent of big data analytics has revolutionized the way we approach carbon removal, allowing for comprehensive performance tracking and trend analysis. By harnessing the power of big data, project teams can identify patterns and correlations that were previously hidden, unlocking opportunities for optimization. Machine learning further enhances this capability, offering predictive insights that guide strategic adjustments. In carbon capture, machine learning algorithms can analyze vast datasets to identify the most effective removal techniques, predict maintenance needs, and optimize energy use. These data-driven insights empower teams to make proactive decisions, ensuring that carbon removal strategies remain cutting-edge and impactful.

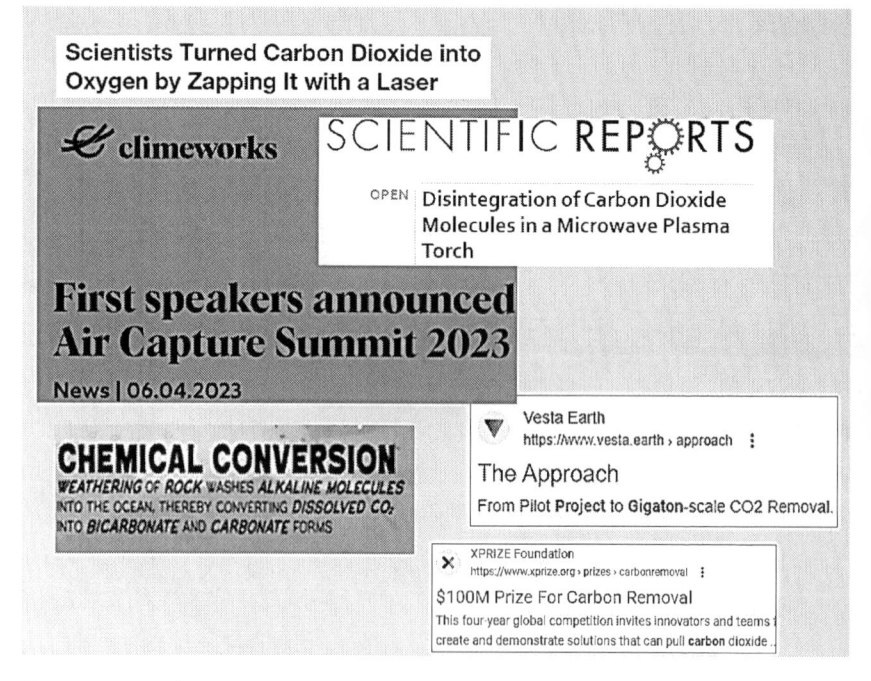

Examples of iterative improvements in carbon projects abound, each showcasing the transformative potential of continuous refinement. Consider the advancements in **algae-based carbon capture**, where successive iterations have led to significant enhancements in efficiency and scalability. Early algae systems faced challenges related to growth rates and carbon absorption capacity. However, through iterative experimentation and process optimization, researchers have developed strains of algae with enhanced carbon uptake, paving the way for large-scale deployment. Similarly, **direct air capture (DAC) technology** has seen remarkable progress through iterative design changes. Initial DAC systems were often hindered by high energy demands and limited capture efficiency. Yet, by applying iterative methodologies, engineers have refined these systems, reducing energy requirements and increasing carbon capture rates. These examples underscore the value of continuous improvement, demonstrating how small, incremental changes can lead to substantial advancements in carbon removal.

12.3 Addressing Unintended Consequences

In the quest to remove carbon dioxide from the atmosphere, we must tread carefully to avoid unintended consequences. Carbon removal strategies are not without their risks, and these can sometimes ripple through ecosystems and communities in unexpected ways. Consider ocean-based sequestration, a promising yet complex method. By iron-seeding algae blooms, we can achieve gigaton-scale carbon dioxide removal. However, this approach risks disrupting marine ecosystems. The sudden influx of nutrients may trigger uncontrolled algae growth, potentially leading to oxygen-deprived zones that can harm aquatic life. Balancing these benefits and risks requires a nuanced understanding of ocean dynamics and a commitment to monitoring and adaptation.

On land, large-scale carbon removal projects can also pose challenges. For example, afforestation and reforestation initiatives, while vital for capturing carbon, can lead to significant changes in land use. Such changes might displace communities or alter local economies, particularly if agricultural land is converted to forests. This is not just a question of environmental Impact; it touches on social justice and community livelihoods. In contrast, offshore wind farms, another tool in the carbon reduction toolkit, usually pose less risk of displacement as they utilize space over open water. Yet, they too, must be carefully sited to minimize impacts on marine life and shipping routes.

To address these potential pitfalls, robust risk assessment and management practices are paramount. Environmental impact assessments (EIAs) are a critical first step, providing a detailed analysis of how a project might affect local ecosystems and communities. These assessments should be comprehensive, considering both direct impacts and potential ripple effects. Scenario planning further aids in this process, allowing project developers to envision a range of possible future outcomes and prepare accordingly. By anticipating various scenarios, we can put in place strategies to mitigate negative impacts before they occur.

Reducing the likelihood and severity of adverse effects involves adopting adaptive management practices. This approach means remaining flexible and responsive, adjusting strategies as new information and conditions arise. Stakeholder engagement is also crucial, ensuring that affected communities have a voice in decision-making processes. By involving local residents, businesses, and environmental groups from the outset, projects can benefit from valuable insights and foster goodwill. Implementation of safeguards and contingency plans provides a safety net, ensuring that unforeseen issues can be addressed swiftly and effectively.

Several projects have successfully navigated the challenges of unintended consequences, offering valuable lessons. Consider adaptive management in afforestation projects, where ongoing monitoring and community involvement have been key. By continuously assessing the health of newly planted forests and adjusting plans as needed, these projects have managed to maintain ecological balance while achieving carbon sequestration goals. Community co-benefit programs in carbon offset initiatives are another example. By providing economic incentives and social benefits to local populations, these programs ensure that carbon removal efforts also support community development.

Offshore wind farms along the East Coast of the United States illustrate a balanced approach to risk management. These projects have worked closely with local stakeholders, including fishermen and environmentalists, to minimize ecological disruptions and optimize site selection. By prioritizing transparency and collaboration, they have successfully mitigated many of the potential negative impacts associated with large-scale renewable energy projects. Such examples demonstrate that with thoughtful planning and execution, it is possible to achieve significant carbon removal while respecting both nature and the communities that depend on it.

12.4 Enhancing Carbon Removal Efficiency

In our quest to effectively combat climate change, enhancing the efficiency of carbon removal strategies is paramount. Innovations in this field are continuously evolving, pushing the boundaries of what's possible. High-efficiency sorbents for direct air capture represent one of the most promising advancements. These materials are designed to absorb CO_2 from the atmosphere with greater efficiency, reducing energy consumption and increasing capture rates. Imagine a sponge, but for carbon dioxide—these sorbents can be reused multiple times, making them both cost-effective and environmentally friendly. The development of these sorbents is not just a technical challenge but a testament to human ingenuity in the face of environmental necessity.

Another groundbreaking approach is enhanced weathering, which accelerates natural mineral carbonation processes. This technique involves spreading finely ground minerals, like olivine or basalt, over large areas to react with CO_2 in the air, forming stable carbonates. The beauty of this process lies in its simplicity and effectiveness, as it mimics natural geological processes that have been occurring for millions of years. By enhancing these processes, we can significantly increase the rate at which carbon is sequestered, offering a scalable solution to the emissions problem. These innovations highlight the creativity and determination driving the carbon removal sector forward.

Collaboration is the cornerstone of enhancing carbon removal efficiency. Public-private partnerships are particularly vital, as they bring together the resources and expertise of government agencies and private companies. These collaborations foster innovation by providing funding, shared knowledge, and risk-sharing opportunities. Academic-industry partnerships also play a crucial role in advancing research and innovation. Universities work alongside businesses to develop cutting-edge technologies, turning theoretical concepts into practical applications. This synergy between academia and industry accelerates the pace of technological advancement, ensuring that the most promising ideas are brought to fruition.

Economic and policy incentives are powerful tools for driving efficiency improvements in carbon removal. Governments can implement tax credits and subsidies for energy-efficient carbon capture technologies, making them more financially attractive to businesses. Such incentives encourage companies to invest in and adopt advanced carbon capture solutions, accelerating their deployment. Policy frameworks that support innovative projects are equally important. These frameworks create an environment that nurtures experimentation and development, providing clear guidelines and support for new technologies. By aligning economic interests with environmental goals, these incentives ensure that efficiency improvements remain a priority in the carbon removal sector.

Several projects serve as exemplars of efficient carbon removal, showcasing the potential of technological advancements and strategic planning. Bioenergy with carbon capture and storage (BECCS) projects have demonstrated remarkable success in capturing and storing CO_2 while generating renewable energy. These projects not only reduce emissions but also contribute to sustainable energy production, exemplifying the dual benefits of integrated carbon removal solutions. Efficient carbon utilization in industrial processes also highlights the potential for significant gains. By capturing and repurposing CO2 emissions from industrial activities, these projects transform waste into valuable resources, reducing overall carbon footprints.

As we explore these advancements, we see the outlines of a future where carbon removal is not only effective but also efficient and economically viable. The innovations and collaborations driving these efforts are paving the way for a more sustainable world. By harnessing the power of technology and strategic partnerships, we can enhance carbon removal efficiency, making meaningful strides toward our climate goals. These efforts are not isolated but part of a larger movement to address the challenges of climate change with creativity, determination, and resilience.

Reflection Section

To bring the concepts discussed in this book to life, consider using a carbon footprint calculator to assess your own impact. Reflect on how changes in your lifestyle could contribute to carbon removal efforts. This exercise not only personalizes the idea of carbon sequestration but also empowers you to take action to reduce your carbon footprint. Explore how small adjustments in daily habits can align with larger global efforts to combat climate change.

Consider reflecting on a carbon removal project or initiative you are familiar with. Identify potential areas for improvement and brainstorm ways to apply iterative methodologies to enhance its effectiveness. This exercise encourages creative thinking and highlights the importance of continuous learning in the pursuit of environmental sustainability.

Conclusion

As we reach the end of our journey exploring the reversal of climate change through gigaton carbon dioxide removal, I hope you've found the insights and strategies discussed both enlightening and empowering. Throughout this book, we've delved into the groundbreaking research of climatologist James Hansen, uncovering the linear relationship between Earth's temperature, CO2 levels, and sea levels. This revelation, akin to an archaeologist unearthing a hidden artifact, has provided us with a roadmap for potential future actions.

We've examined the feasibility of large-scale carbon removal technologies, the power of community-driven initiatives, and the crucial role of billionaire-led innovations in achieving our climate goals. The path ahead is not without challenges, but the success stories we've explored demonstrate that change is possible when we work together.

Imagine a future where our collective efforts have led to significant milestones in gigaton carbon dioxide removal. Picture a world where direct air capture facilities dot the landscape, efficiently removing CO2 from the atmosphere. Envision vast swathes of land restored through reforestation and regenerative agriculture, creating vibrant ecosystems that not only sequester carbon but also support biodiversity. These scenarios are not mere fantasies; they are possibilities within our reach.

As we look to the future, it's crucial to consider both the possibilities and the probabilities. While the road ahead may seem daunting, I firmly believe that we have the knowledge, tools, and collective will to make a difference. By embracing innovative technologies, supporting community-driven initiatives, and fostering collaboration across sectors, we can tip the scales in favor of a sustainable future.

But the journey doesn't end here. As you close this book, I encourage you to reflect on your own role in this global effort. Every action counts, no matter how small it may seem. Whether it's making sustainable lifestyle choices, supporting climate-friendly policies, or engaging in local advocacy, talking with your Senator and Congressman, your voice and your actions have the power to create ripples of change. This is especially important when, as in this case, most people have not been informed about the necessity of CO_2 removal from the heat blanket above. It should be allowing infrared (IR) radiation to reflect back out into space and not become trapped in the surface layers of our air. CO_2 is a very potent IR absorber.

The reversal of climate change is not a task for one individual or one nation alone. It requires a collective awakening, a shared commitment to healing our planet. By working together, we can build resilience, adapt to the challenges ahead, and create a world where both nature and humanity can thrive.

As a physicist and professional engineer with over three decades of experience, I've witnessed firsthand the transformative power of innovation and collaboration. From my work on electrotherapy devices to my research on zero-point energy, I've seen how seemingly impossible challenges can be overcome when we approach them with curiosity, determination, and an open mind.

So, as you embark on your own journey to combat climate change, remember that you are part of a global community united by a common purpose. Together, we can reverse the tide of global warming and build a sustainable future for generations to come. The path ahead may be uncertain, but one thing is clear: the time for action is now. Let us seize this moment, armed with the knowledge and tools we need to make a difference. If you have connections to billionaires or the few emerging trillionaires, let's talk about proposals with milestones to ramp up Gigaton CDR as soon as possible. The 50-page **McKinsey Report** *"Carbon removals: How to scale a new gigaton industry"* is a great document to share with potential benefactors.

Let us be the architects of a brighter, more sustainable tomorrow. Together, we can turn the hopeful possibilities of gigaton carbon dioxide removal into a probable reality. The power to shape our future lies in our hands. Let us wield it wisely, with compassion and determination, for the sake of our planet and all who call it home.

> THE "BOILING FROG" STORY IS A METAPHOR SUGGESTING THAT IF A FROG IS PLACED IN BOILING WATER, IT WILL JUMP OUT, BUT IF PLACED IN TEPID WATER THAT IS SLOWLY HEATED, IT WON'T PERCEIVE THE DANGER AND WILL BE BOILED ALIVE, HIGHLIGHTING HOW PEOPLE CAN BECOME ACCUSTOMED TO DANGEROUS SITUATIONS THAT DEVELOP GRADUALLY.
>
> — Google

References

- Hansen 2006: *The threat to the planet - GISS Publications*
  https://pubs.giss.nasa.gov/abs/ha06110e.html
- *CO2 lags temperature rises in ice core data*
  http://hyperphysics.phy-astr.gsu.edu/hbase/thermo/icecore.html
- *Carbon removals: How to scale a new gigaton industry*
  https://www.mckinsey.com/capabilities/sustainability/our-insights/carbon-removals-how-to-scale-a-new-gigaton-industry
- *Carbon Dioxide | Vital Signs – Climate Change - NASA*
  https://climate.nasa.gov/vital-signs/carbon-dioxide/
- *The Messenger* by James Hansen, *Technology Review*,
  https://www.technologyreview.com/2006/07/01/228690/the-messenger/
- *Analysis of Vostok Ice Core Data*
  https://globalchange.umich.edu/globalchange1/current/labs/Lab10_Vostok/Vostok.htm
- Hansen, J. (1988). Global Climate Changes as Forecast by Goddard Institute for Space Studies Three-Dimensional Model. *Journal of Geophysical Research*, 93, 9341-9364.
  https://doi.org/10.1029/JD093iD08p09341
- *Global warming in the pipeline | Oxford Open Climate Change*
  https://academic.oup.com/oocc/article/3/1/kgad008/7335889
- *Carbon removals: How to scale a new gigaton industry*
  https://www.mckinsey.com/capabilities/sustainability/our-

- insights/carbon-removals-how-to-scale-a-new-gigaton-industry
- *Global change over the last climatic cycle from Vostok...* https://www.sciencedirect.com/science/article/pii/1040618289900177
- *Carbon removals:How to scale a new gigaton industry* https://www.mckinsey.com/capabilities/sustainability/our-insights/carbon-removals-how-to-scale-a-new-gigaton-industry
- *CLIMATE CHANGE 2023* https://www.ipcc.ch/report/ar6/syr/downloads/report/IPCC_AR6_SYR_FullVolume.pdf
- *Hothouse Earth: An Inhabitant's Guide,* Bill McGuire, Icon Books, UK, 2022, also by Penguin Books India, https://read.amazon.com/?asin=B0B1THV9DQ&ref_=dbs_t_r_kcr
- *Elon Musk -- why 'all energy generation will be solar'* https://m.economictimes.com/industry/renewables/elon-musk-on-why-all-energy-generation-will-be-solar/articleshow/113723423.cms
- *The Bezos Earth Fund has pumped billions into climate ...* https://www.theguardian.com/environment/article/2024/may/20/jeff-bezos-earth-fund-carbon-offsets-climate-sector-uneasy-aoe
- *Bill Gates firm gives $40 mln for Canadian carbon tech test ...* https://www.reuters.com/sustainability/climate-energy/bill-gates-firm-gives-40-mln-canadian-carbon-tech-test-bed-2024-12-18/
- *Virgin Earth Challenge - Wikipedia* https://en.wikipedia.org/wiki/Virgin_Earth_Challenge#:~:text

=The%20Virgin%20Earth%20Challenge%20was,materially%20in%20global%20warming%20avoidance.

- *Top 20 Direct Air Capture Companies In 2023*
https://carbonherald.com/top-20-direct-air-capture-companies/

- *A Research Strategy for Ocean Carbon Dioxide Removal ...*
https://www.nationalacademies.org/our-work/a-research-strategy-for-ocean-carbon-dioxide-removal-and-sequestration

- *Deployment of BECCUS value chains in the United States*
https://www.ieabioenergy.com/wp-content/uploads/2023/03/BECCUS-1.0_US-Case-Study_final_update.pdf

- *The CarbFix Pilot Project–Storing carbon dioxide in basalt*
https://www.sciencedirect.com/science/article/pii/S1876610211008253

- *The Sunrise Movement: how a US grassroots youth ...*
https://rapidtransition.org/stories/the-sunrise-movement-how-a-us-grassroots-youth-movement-helped-set-the-national-climate-agenda-for-rapid-change/

- *Copenhagen's Journey Towards Carbon Neutrality*
https://www.linkedin.com/pulse/case-study-copenhagens-journey-towards-carbon-neutrality-yigsf

- *Indigenous Guardians of Global Biodiversity and Climate at ...*
https://amazonwatch.org/news/2024/1114-guardians-of-global-biodiversity-and-climate-uplifting-indigenous-solutions-at-cop16

- *Benefits of Green Infrastructure | US EPA*
https://www.epa.gov/green-infrastructure/benefits-green-infrastructure

- Hansen et al. 2023: Global warming in the pipeline - Pubs. GISS https://pubs.giss.nasa.gov/abs/ha09020b.html
- Global Climate Agreements: Successes and Failures https://www.cfr.org/backgrounder/paris-global-climate-change-agreements
- The Environmental Justice Movement https://www.nrdc.org/stories/environmental-justice-movement
- Carbon Footprint Calculator | US EPA https://www.epa.gov/ghgemissions/carbon-footprint-calculator
- 12-Ways to Live More Sustainably https://www.biologicaldiversity.org/programs/population_and_sustainability/sustainability/live_more_sustainably.html
- How You Can Influence Local Environmental Policies https://www.birdhouse.farm/rna-blog/advocacy-in-action-how-you-can-influence-local-environmental-policies
- Climate Literacy and Energy Awareness Network (CLEAN) https://ceee.colorado.edu/programs/climate-literacy-and-energy-awareness-network-clean
- Nations must go further than current Paris pledges or face ... https://www.unep.org/news-and-stories/press-release/nations-must-go-further-current-paris-pledges-or-face-global-warming
- EUROPEAN UNION: AN EMISSIONS TRADING CASE ... https://www.edf.org/sites/default/files/eu-case-study-may2015.pdf
- The Global South has mastered locally-led climate ... https://climatepromise.undp.org/news-and-stories/global-

south-have-mastered-locally-led-climate-adaptation-solutions-its-time-scale

- UN Climate Change 2023 Highlights https://unfccc.int/about-us/2023-highlights

- Climate crisis costs 12% in GDP for each 1°C temperature rise https://www.weforum.org/stories/2024/06/nature-climate-news-global-warming-hurricanes/#:~:text=Climate%20change%20costs%20the%20world,times%20larger%20than%20previous%20estimates.

- The new analysis outlines national opportunities to remove ... https://www.llnl.gov/article/50686/new-analysis-outlines-national-opportunities-remove-carbon-dioxide-gigaton-scale

- Taking power as individuals (and why individual climate ... https://www.brookings.edu/articles/taking-power-as-individuals-and-why-individual-climate-action-cant-save-us/

- Effective Strategies to Mitigate Eco-Anxiety: Turning Fear ... https://www.climatize.earth/effective-strategies-to-mitigate-eco-anxiety-turning-fear-into-action-for-a-sustainable-future/

- Climate communication: 10 research-backed tips | Resources https://hsph.harvard.edu/research/health-communication/resources/climate-communication-tips/

- Case Studies - Climate Justice https://www.mrfcj.org/our-work/case-studies/

- US Climate Alliance https://usclimatealliance.org/

- James E. Hansen | Climate Science, Awareness and Solutions https://csas.earth.columbia.edu/about/people/james-e-hansen

- *Getting into the details of carbon accounting* https://www.llnl.gov/article/51821/getting-details-carbon-accounting#:~:text=At%20the%20most%20fundamental%20level,of%20a%20trusted%20CDR%20industry.

- *Carbon capture and storage: What can we learn from the ...* https://www.catf.us/resource/carbon-capture-storage-what-can-learn-from-project-track-record/

- *Direct Air Capture - Energy System - IEA* https://www.iea.org/energy-system/carbon-capture-utilisation-and-storage/direct-air-capture

- Valone, T. 2016. *Nikola Tesla's Electricity Unplugged: Wireless Transmission of Power as the Master of Lightning Intended.* Kempton, IL: Adventures Unlimited Press.

- Valone, T. 2020. *The Future of Energy: Challenges, Perspectives, and Solutions.* Hauppauge, NY: Nova Science Publishers, Inc.

- Werner, M., and T. Stockli. 2012. *Life from Light: Is it possible to live without food? A scientist reports on his experiences.* East Sussex, UK: Clairview Books Ltd.

- Lynas, M. 2020. *Our Final Warning: Six Degrees of Climate Emergency,* 4th Estate, An Imprint of HarperCollinsPublishers, UK.

- Maor, R. 2018. *A Year Without Food: Discover the Unimaginable World of Proven Energetic Nourishment.* Aingeal Rose & Ahonu, designed and eds. Prescott, AZ: Twin Flame Productions LLC.

- Renner, J. L. 2006. "*The Future of Geothermal Energy.*" United States. . Accessed on June 27, 2022.

- Tenning, M. 2015. "*Water, Water Everywhere.*" Scientific American 313 (3): 26.

- GISTEMP Team. 2022. *GISS Surface Temperature Analysis (GISTEMP), version 4*. NASA Goddard Institute for Space Studies. Dataset accessed 2022-06-14.
- Lenssen, N., G. Schmidt, J. Hansen, M. Menne, A. Persin, R. Ruedy, and D. Zyss. 2019. "Improvements in the GISTEMP Uncertainty Model." *JGR: Atmospheres* 124 (12): 6307–6326, .
- Lord, J., A. Thomas, N. Treat, M. Forkin, R. Bain, P. Dulac, C. H. Behroozi, T. Mamutov, J. Fongheiser, N. Kobilansky, S. Washburn, C. Truesdell, C. Lee, and P. H. Schmaelzle. 2021. "Global potential for harvesting drinking water from air using solar energy." *Nature* 598: 611–617.
- Valone, Thomas F., "Predictive Connection for 2100 between Atmospheric Carbon, Global Warming and Ocean Height Based on Climate History" *International Journal of Environment and Climate Change*, 2019, Vol. 9, Issue 10, p. 562-593, DOI: 10.9734/ijecc/2019/v9i1030140 – Open Access: https://tinyurl.com/PredictWarming
- Valone, Thomas F., "Linear Global Temperature Correlation to Carbon Dioxide Level, Sea Level, and Innovative Solutions to a Projected 6°C Warming by 2100", *Journal of Geoscience and Environment Protection*, Vol.9 No.3, March 2021, 17,000 views, **DOI:** 10.4236/gep.2021.93007, Open Access: https://tinyurl.com/GlobalTempCO2
- Valone, Thomas F., "Study of a Possible Global Environmental Forecast and Roadmap Based on 420 kY of Paleoclimatology", *Modern Advances in Geography, Environment and Earth Sciences*, Vol. 5, Publisher: B P International, July 2021, Chapter, DOI: 10.9734/bpi/magees/v5/2845F, https://tinyurl.com/climateforecast
- Valone, Thomas, "Predictive Connection of CO2, Warming, and Sea Level", Conference on Future Energy (COFE12) 2020 and IEEE-ISTAS, Tufts University, Boston MA, Nov. 15-16, 2019,

Content in special issue of Advances in Science, Technology, and Engineering Systems Journal (ASTES), as well as in the Journal of Atmospheric Research. Video slideshow online at: https://tinyurl.com/ClimateVideoValone.

- Hernsath, Klaus. *Climate Change Reversal: Restoring Earth for Future Generations*. Outskirts Press 2015 paperback and Kindle

- Mildenberger, Matto. *Carbon Captured: How business and labor control climate politics*, The MIT Press, 2020

- McLaren, Duncan & Olaf Corry, "Carbon Dioxide Removal: What is Sustainable and Just?", *Environment: Science and Policy for Sustainable Development*, Vol. 67, Issue 1, Feb. 2025. Open access.

- Broome, John. *Climate Matters: Ethics in a Warming World*, Norton & Company, paperback 2012

- Gates, Bill. *How to Avoid a Climate Disaster: The Solutions We Have and the Breakthroughs We Need*, Vintage Books, 2012 (New York Times Best Seller)

- Hawken, Paul, Editor. *Drawdown: The most comprehensive plan ever proposed to reverse global warming*, Penguin Books, large-size paperback, 2017 (New York Times Best Seller)

- Keeling, C. D., Bacastow, R. B., Bainbridge, A. E., Ekdahl, C. A., Guenther, P. R., Waterman, L. S., and Chin, J. F. S.: Atmospheric carbon dioxide variations at Mauna Loa Observatory, Hawaii, Tellus A., 28, 538–551, https://doi.org/10.1111/j.2153-3490.1976.tb00701.x, 1976.

- Keeling, R. F. and Manning, A. C.: 5.15 – Studies of Recent Changes in Atmospheric O2 Content, in: Treatise on Geochemistry, 2nd Edn., edited by: Holland, H. D. and

Turekian, K. K., Elsevier, Oxford, 385–404, https://doi.org/10.1016/B978-0-08-095975-7.00420-4 , 2014

- Goodell, Jeff. *How to Cool the Planet: Geoengineering Audacious Quest to Fix Earth's Climate*, www.hmhbooks.com, 2010

- Thunberg, Greta. The Climate Book, Penguin Random House, 2022, www.penguinrandomhouse.com

Made in the USA
Columbia, SC
29 March 2025